THE NEW NATURALIST

A SURVEY OF BRITISH NATURAL HISTORY

BRITAIN'S STRUCTURE AND SCENERY

The aim of this series is to interest the general reader in the wild life of Britain by recapturing the inquiring spirit of the old naturalists. The Editors believe that the natural pride of the British public in the native fauna and flora, to which must be added concern for their conservation, is best fostered by maintaining a high standard of accuracy combined with clarity of exposition in presenting the results of modern scientific research. The plants and animals are described in relation to their homes and habitats and are portrayed in the full beauty of their natural colours, by the latest methods of colour photography and reproduction.

THE NEW NATURALIST

BRITAIN'S STRUCTURE AND SCENERY

by

L. DUDLEY STAMP

C.B.E. D.Litt. D.Sc. LL.D. Ekon.D.

PROFESSOR EMERITUS OF SOCIAL GEOGRAPHY
IN THE UNIVERSITY OF LONDON
AT THE LONDON SCHOOL OF ECONOMICS

WITH 47 COLOUR PHOTOGRAPHS
40 PHOTOGRAPHS IN BLACK AND WHITE
74 MAPS AND DIAGRAMS

COLLINS ST JAMES'S PLACE LONDON

First Published in September 1946 by
Collins, 14 St James's Place, London
Produced in conjunction with Adprint
and printed in Great Britain
by Collins Clear-Type Press
London and Glasgow

Second Edition 1947
Third Edition 1949
Fourth Edition 1955
Fifth Edition 1960
Sixth Edition 1967

TO

A MUCH LAMENTED LOVER OF THE COUNTRYSIDE

THE LATE LORD JUSTICE SCOTT, P.C.

TO SERVE UNDER WHOM FOR A YEAR

WAS A LIBERAL EDUCATION

CONTENTS

CHAPTER		PAGE
	Introduction	1
1	Highland Britain and Lowland Britain	5
2	Reading the Rocks	10
3	Earth History—Time and Life	18
4	The Geological Map	30
5	Land Forms and Scenery—the Work of Rivers	37
6	The Work of the Sea	51
7	The Scenery of the Sedimentary Rocks	62
8	The Scenery of Limestone Country	73
9	The Land Forms of Volcanic Country	77
10	The Scenery of Glaciation	82
11	Soils	91
12	Geographical Evolution	103
13	The Pliocene Period	147
14	The Great Ice Age and After	154
15	The Regions of Britain—The London and Hampshire Basins	172
16	The Weald	184
17	East Anglia and the Fens	193
18	The English Scarplands	200
19	The English Midlands	207
20	The South-West	211
21	The Welsh Massif	217
22	The North of England—the Lakes and the Pennines	222
23	Scotland	227
24	Ireland	236
	Annotated Bibliography	241
	Index	245

LIST OF COLOUR PLATES

FACING PAGE

1A A Fault, Swanbridge, Glamorganshire Coast 4
1B Stac Polly, Wester Ross 4
2 Autumn in Aberglaslyn, North Wales 5
3A Culm Measure Sandstones at Bude, Cornwall 12
3B Vertical Sea Cliffs at Crackington Haven, Cornwall 12
4A Cotton-grass moor near the West Country Inn, Hartland, Devon 13
4B Vertical strata, Compass Point, Bude, Cornwall 13
5A The Coast of South Pembrokeshire 64
5B Flat-topped hills near Honiton, Devon 64
6 Cheddar Gorge, Somerset 65
7A Ulpha Fell, Lake District 80
7B Chalk Cliffs, Bempton, Yorkshire 80
8A Llyn Du'r-arddu, Snowdonia 81
8B A coastal scree of Sandstone and Shale, Cornwall 81
9A Winter in Fenland 84
9B Soil-Creep, Widemouth Bay, North Cornwall 84
10 A typical Podsol developed on current-bedded Lower Greensand 85
11 A black, peaty meadow soil, overlying chalk, near South Moreton, Berkshire 92
12 Colour photograph of a model of the southern part of the Cumberland Coalfield 93
13 Blaen-Rhondda Colliery, South Wales 128
14 Opencast Coalworking at Waunavon, South Wales 129
15A A Coal Measure Forest. From a diorama in the Geological Museum 144
15B A Jurassic Landscape. From a painting in the Geological Museum 144
16A A Landscape of London Clay Times. From a diorama in the Geological Museum 145
16B The Thames Valley in the Old Stone Age. From a diorama in the Geological Museum 145

FACING PAGE

17 The Malvern Hills from the north-east 148
18 The Pass of Llanberis, North Wales 149
19 Mosedale in September (near Wasdale, Lake District) 156
20 The Marshes of the Afon Glaslyn, looking north from Portmadoc 157
21 The Chalk Scarp of the North Downs, looking eastwards from near Box Hill, Surrey 180
22 Iping Common, near Midhurst, Sussex 181
23 A typical dry Chalk Valley. Tappington Farm, near Denton, Kent 188
24 The Scarp of the Cotswolds, near Birdlip Hill, Gloucestershire 189
25 Dartmoor from Hatherleigh Moor, Devon 196
26A Carrick Roads, looking south from near Penpoll, Truro 197
26B The Truro River at Malpas, Cornwall 197
27A The moorland above the Rhondda Valley 204
27B Cwm-parc and Parc Colliery, the Rhondda 204
28A Near Nevin, North Wales 205
28B Gateholm, Pembrokeshire 205
29 Snowdon from the east, with Llynau Mymbyr in the foreground 228
30A The Scafell Range from the summit of Great Gable 229
30B The Screes, Wastwater 229
31A Millstone Grit Scarp, north-west of Hebden Bridge, Yorkshire 236
31B Looking west to Crummock Water 236
32A Stac an Armin, Stac Lee, and Boreray, St. Kilda, from the south-west 237
32B Stac Polly from Loch Lurgainn, Wester Ross 237

LIST OF PLATES IN BLACK AND WHITE

FACING PAGE

I Langsleddale, Lake District 32
II The Chalk Coast of Dorset, with Old Harry Rock 33
III (a) The Whale Rock, Bude, North Cornwall 36
III (b) Folded Rocks exposed on the Foreshore, north of Bude 36
IV (a) Strata Cliff, Millook, North Cornwall 37
IV (b) An Unconformity at Horton-in-Ribblesdale, Yorkshire 37
V The Landslip, St. Catherine's Point, Isle of Wight 44
VI Honeycomb Weathering of Sandstone 45
VII The Rocky Bed of the River Usk, near Brecon 48
VIII (a) Details of a Pothole, River Usk, near Brecon 49
VIII (b) Head of Wastwater, Lake District 49
IX The Coast at Folkestone, Kent, looking eastwards 96
X Blakeney Point, Norfolk, looking east 97
XI A Pre-Glacial Rock Platform, North Coast of Islay 100
XII The Culbin Sand Dunes, Morayshire 101
XIII Pipes in Chalk, near Canterbury, Kent 108
XIV The Dripping Well, near Knaresborough, Yorks 109
XV (a) Fingal's Cave, Isle of Staffa 112
XV (b) Duncryne, Gartocharn, Dumbartonshire 112
XVI (a) Arthur's Seat, Edinburgh 113
XVI (b) Striding Edge, Helvellyn, Lake District 113
XVII The Aletsch Glacier, Switzerland 160
XVIII Soil Erosion in England 161
XIX The Needles, Isle of Wight, looking east 164
XX (a) Portland Bill, Dorset (Portland Limestone) 165
XX (b) South-west of St. David's, Pembrokeshire 165
XXI (a) Near Clifden, Connemara, Ireland 172

		FACING PAGE
XXI (b)	Remains of a Pine Forest in Peat, near Daless, Nairn	172
XXII (a)	Contorted Gravels of the River Cam, Great Shelford, Cambridgeshire	173
XXII (b)	The Lead Hills, Southern Uplands	173
XXIII	The Norfolk Broads	176
XXIV	Salt Marshes on the Essex Coast	177
XXV	Chichester Harbour, Sussex	208
XXVI	Lulworth Cove, Dorset	209
XXVII	A Tor on Dartmoor	212
XXVIII	Rocky Valley, Tintagel	213
XXIX	Llyn-y-Cae, Cader Idris, North Wales	220
XXX	Giggleswick Scar	221
XXXI	Successive Alluvial Terraces of the River Findhorn, Scotland	224
XXXII	Stac Lee, St. Kilda	225

Every care has been taken by the Editors to ensure the scientific accuracy of factual statements in these volumes, but the sole responsibility for the interpretation of facts rests with the Authors.

LIST OF ILLUSTRATIONS IN THE TEXT

FIGURE PAGE

1 Highland Britain and Lowland Britain 6
2 The Mean Annual Rainfall of the British Isles 8
3 The Chief Moorlands of Great Britain 9
4 The Geological Column 15
5 Diagrams illustrating Exfoliation 20
6 Diagram of the Fault shown in Plate 1A 23
7 Diagrammatic Section of an Unconformity 24
8 Major Episodes in Earth History 26
9 Some Characteristic Fossils 27
10 Diagram illustrating the distribution in time and space of a typical fossil 29
11 A Simplified Geological Map of the British Isles 34-35
12 Diagram of a Meandering River 40
13 Diagrammatic Section through the Deposits of a Delta and a Lake 41
14-16 Stages in the Development of the Drainage of the Weald 42-43
17 Diagrammatic Sections along a *Talweg* 47
18 Sections showing the formation of Cliffs 52
19 Section through a Raised Beach 53
20 Diagrams showing the Drift of Shingle along a Shelving Beach 55
21 A Shingle Spit 56
22 An Atlantic Coastline 58
23 A Drowned Coastline 60
24 Diagrammatic Explanation of the Peneplanation shown in Plate 5A 64
25 Diagrammatic Explanation of the Flat-topped Hills of Plate 5B 64
26 Map of the Scarplands of England 66
27 Forms of Cuestas, showing relation of surface form to dip 67

FIGURE		PAGE
28	Sand Grains under the Microscope	69
29	Section through an Anticline and two Synclines	70
30	Sections showing the effects of an asymmetric syncline or dipping strata on the shift of a Valley	71
31	Section through a Batholith showing the Metamorphic Aureole	78
32	Section through a Laccolite and a Phacolite	79
33	Section through a Volcano	80
34	Sections showing Crag-and-Tail Structure	85
35	Diagram of an Outwash Fan	86
36	Map of the Cirques in the neighbourhood of Snowdon	88
37	Diagram of the Podsol overlying Current-bedded Sands shown in Plate 10	93
38	The Geography of Cambrian Times	106
39	The Geography of Ordovician Times	107
40	The Geography of Silurian Times	110
41	Caledonian Folding and the Geography of Old Red Sandstone Times	113
42	The Geography of Early Carboniferous Limestone Times	114
43	The Geography of Late Carboniferous Limestone Times	115
44	The Geography of Millstone Grit Times	118
45	The Geography of Coal Measure Times	122
46	Armorican Folding and the Geography of Permian Times	124
47	The Geography of Magnesian Limestone Times	125
48	The Geography of Bunter Sandstone Times	126
49	The Geography of Keuper Times	127
50	The Geography of Liassic Times	130
51	The Geography of Middle Jurassic Times	132
52	The Geography of Upper Jurassic Times	134
53	The Geography of Lower Cretaceous Times	137

FIGURE PAGE
54 The Geography of Upper Cretaceous Times 139
55 The Geography of Eocene Times 141
56 Cycles of Sedimentation in the English Tertiary 143
57 Alpine Folding and Vulcanicity 144
58 The Geography of Early Pliocene Times 150
59 The Late Pliocene Chillesford River 152
60 The Geography of the Period of Maximum Glaciation 160
61 Glacial Lakes in the Midlands and North of England 166
62 The Regions of Britain 174
63 The Structural Elements in the Geography of Britain 175
64 Section across the London Basin 179
65 Section across the Thames River Terraces 179
66 Section across the Hampshire Basin and the Isle of Wight 182
67 Section across the Weald of Kent and Sussex 185
68 Diagram of the Scarp of the North Downs shown in Plate 21 187
69 Diagram showing the Origin of the Drainage of South-eastern England 201
70 Diagram of the Scarp of the Cotswolds shown in Plate 24 202
71 Diagram showing the formation of a cirque 218
72 Section through the Rift Valley of Scotland 230
73 Section through the North-West Highlands of Scotland 232
74 Diagrammatic Map showing the Structure of Ireland 238

EDITORS' PREFACE

IT IS ONE of the principal objects of the NEW NATURALIST series to present in simple language to the lay reader the results of recent scientific work in the many fields covered by the general term "Natural History." Another is to take the results of laboratory research into the realm of field studies and particularly to recapture the spirit of the old naturalists whose keen delight was in the study of animals and plants in their native haunts.

The present volume may be regarded in many respects as a background volume to the whole series in that it attempts to trace the evolution, through the many millions of years of geological time, of the geography of the British Isles and so to present a general view of the stage and setting of Britain's Natural History.

The task has been rendered especially difficult for several reasons. In the first place it has been necessary to compress a large section of the science of geology into a very small space ; in the second place it has been necessary to eliminate a whole scientific terminology which to the geologist makes for brevity and precision but which would be unfamiliar to the non-geologist. In addition, any attempt to reconstruct the geography of past ages is beset with pitfalls, so that the generalisations here presented may appear to have a definiteness which is not warranted by the facts. They must be regarded as liable to constant revision and even now, as the results of the borings undertaken in the intensive war-time search for oil are studied, they may be greatly modified.

THE EDITORS

ACKNOWLEDGMENTS

Some of the diagrams in this book have been based on ones illustrating my own previously-published works—in particular *An Introduction to Stratigraphy* (Allen & Unwin) and *The British Isles, A Geographic and Economic Survey* (Longmans)—but in all cases have been modified, often to include recent advances in knowledge.

In the provision of a number of photographs I am much indebted to H.M. Geological Survey. The simplified geological map (Fig. 11) is based on those published by the Survey.

L.D.S.

NOTE TO THE SECOND EDITION

My sincere thanks are due to many correspondents for their helpful comments, but especially to my friends, Dr. S. Graham Brade-Birks and Dr. D. A. Osmond.

December, 1946. L.D.S.

NOTE TO THE FOURTH EDITION

The opportunity has been taken to make a few minor corrections and to add some recent references. Since this book first appeared a number of the volumes in the *New Naturalist Series* have extended and amplified many of the topics here covered very briefly. I deal here with the evolution of our scenery through the aeons of geological time: I have now carried on my story by tracing the hand of man up to the present day in *Man and the Land* recently issued in the *New Naturalist Series*.

October, 1954. L.D.S.

NOTE TO THE SIXTH EDITION

Once again I am grateful to friends who have called my attention to misprints and to statements calling for modification, but all changes in this edition are of a minor character.

June, 1966. L.D.S.

INTRODUCTION

THE WEALTH of a country's fauna and flora is not to be measured by numbers of species alone. Its wealth lies rather in variety, and to a naturalist in the British Isles the fascination of the native fauna and flora is in the great variety to be found in a small space. Gilbert White's immortal *Natural History of Selborne* is, in essence, the natural history of a single parish of a few square miles. Yet like many another English parish Selborne, at the western end of the Weald in Hampshire near the borders of Surrey and Sussex, embraces within its limited area many distinctly different habitats or environments, each with its characteristic and often contrasted plants and animals. On the one side lie the open, wind-swept chalk downs with their calcareous soils and lime-loving plants, on the other the coarse sands of the Lower Greensand formation with sterile, acid, hungry soils—too " hungry " to attract the farmer and so given over to heathland and woodland of oak, birch and pine—whilst between the two are the Gault vale with its heavy clay soils and the magnificent " foxmould " developed on the Upper Greensand and accounted one of the finest agricultural soils in the whole of Britain. Such contrasts within a single parish or group of parishes are by no means unusual—indeed parish boundaries were often drawn originally so as to include as great a variety as possible of types of land—and they are reflected in the relief or form of the ground, in soils, in the natural vegetation cover and its associated animal life as well as in the way man, though kept within certain limits, has adapted the natural environment to his own ends. Small differences of elevation, slope, aspect and shelter cause purely local variations in the climate giving rise to different " microclimates " in the area, but they are variations sufficient to spell success or failure in many a farming enterprise, just as they permit or prevent the survival of a given species of the wild flora or fauna.

Those who are accustomed to larger spheres are apt to be obsessed with the discovery that it is by no means difficult to travel by road or rail from coast to coast of Britain, from east to west or even from north to south, in a single day. Yet in 25 miles of such a journey may be found a variety of scenery only to be equalled in a journey of ten times that distance in other lands. The kaleidoscopic rapidity with which the British scene changes is well illustrated from our coastline. It has recently been calculated that the coastline of England and Wales alone —the deeply indented and island-fringed coastline of Scotland is much longer in proportion—is some 2751 miles in length. Within that length may be found mud-flats, sand-dunes, shingle beaches, raised beaches and drowned valleys, sheltered bays and stormy headlands, together with cliffs of the most varied types. The cliffs alone range from the crumbling or slippery boulder clay slopes but a few feet high along parts of the east coast to giants rising almost sheer for a thousand feet from the sea below ; in colour and material they range from the dazzling white chalk of the south and east, or the brilliant red of the New Red Sandstone of Torbay, to the majestic greys and ochres of the granite coasts or the forbidding grey and black of some of the slate cliffs. Inland the story is the same. In the Scottish Highlands it is easy to find a dozen square miles not only without a human habitation or track, but also where the foot of man rarely treads and which even scarcely knows the foot of one of his domestic animals. Yet another dozen square miles of a part of the English lowland may be almost as densely peopled as any similar area on the earth's surface and one where wild nature seems almost to have been eliminated.

England, Wales and Scotland are divided into 85 counties or shires (England 40, Wales 12, Scotland 33) most of which have persisted with few changes of boundary for more than a thousand years. These years have seen vast changes and some of the counties appear to-day anomalous in a modern world—small, sparsely populated, and poor—but, notwithstanding, intensely proud of their history and tradition and jealous of their ancient rights and privileges. Others have become, with vastly increased populations, too unwieldy and have been divided, so that there are now in England 50 Administrative Counties, making a total for Britain, without the Isle of Man, of 95 Administrative Counties, not including the large towns and cities which are County Boroughs having the status of counties. It would be interesting to know how many of the 50,000,000 people of Britain can lay claim to have set foot in each of the 95 counties. There are whole counties so far off the beaten track that not a man in a thousand has visited them or knows anything of the conditions of life in them, often so greatly different from his home area. How many, for example, know the Shetlands where the midwinter sun at noon does not rise more than six degrees above the horizon but where it is possible to read at midnight in summer without artificial light? If one liked to make the test more severe and ask how many of the 50,000,000 have visited every one of the inhabited islands which make up the British Isles it is most probable that the answer would be—none.

All this is intended to suggest the remoteness, the inaccessibility and the generally unknown character of so many parts of what, if one is content to think merely in terms of square miles and average density of population, is a small and very densely peopled country. So much of it is, indeed, a *terra incognita* even to the well-travelled minority. To the new naturalist there is much to be explored. It may be that to find a new species of plant or animal not yet described would be an event of unlikely occurrence, but there are many areas where the vegetation has never been mapped or described, where the changing balance of plant and animal life is waiting to be observed and recorded for the first time and where the explanation of observed changes is still a matter of guesswork. Some of the unexplained features are matters of the highest economic importance—the changing character of hill pastures, the new plant relationships created by afforestation and the introduction of foreign trees are some that spring to mind—and all are pregnant with possible scientific results.

Such are the opportunities awaiting the field observer. He has in his homeland what is in many respects a museum model illustrating the evolution of the world as a whole. For the great variety of environments is the outward and visible reflection of a long and complex geological history. Each of the great ages in the earth's evolution has left its mark on these islands; rocks laid down in all the great periods of geological time are to be found represented in the British Isles.

It thus becomes the purpose of this book to trace the geological evolution of our homeland—to trace, step by step, the building of the British Isles. By this means we are able to understand the structure or the build of its contrasted regions. We are, in fact, attempting to understand the structure and the development of the stage upon which the drama of British natural history is played. In studying the broader aspects we need to consider the British Isles as a whole but, since so many of the books in this series consider primarily the island of Britain—that is England, Wales and Scotland—with its associated smaller islands, we shall consider in more detail and draw most of our examples from it. In tracing the evolution of the structure of the country and of its physical features we are, in fact, attempting to visualise the long history which lies behind the basal elements in its scenery—mountain and plain, hill and dale—recognising at the same time that the intimate details of that scenery are the work of man and lie outside the scope of this volume.

P. L. EMERY

A fault, Swanbridge, Glamorganshire Coast
For explanation see Fig. 6, page 23

J. FISHER

Stac Polly, Wester Ross
Current-bedded red Torridonian Sandstone with well-developed jointing (See page 63 and page 234)

M. WIGHT

Autumn in Aberglaslyn, North Wales (See page 39)

HIGHLAND BRITAIN AND LOWLAND BRITAIN

SIR HALFORD MACKINDER in his now classic book, *Britain and the British Seas*, published in 1902, made a simple yet fundamental distinction between two roughly equal halves of the island of Britain. If one draws a line approximately from the mouth of the Tees to the mouth of the Exe it will be found that all the main hill masses and mountains lie to the north and west, the major stretches of plain and lowland to the south and east. To the north and west lies Highland Britain, to the south and east lies Lowland Britain. There is rarely, in nature, a sharp line between two such regions but rather does the one fade gradually into the other. This is true in Britain; nevertheless it is possible to draw a line with some accuracy and this has been done in Fig. 1.

In *Highland Britain* the dominant character of the country is upland. There are large and continuous stretches more than a thousand feet above sea level; plains and valleys occur but they are of limited extent and tend to form interruptions in the general upland character of the country as a whole. In some places are rugged mountains and even at lower levels crags of rock may appear at the surface. Even where the rocks do not thus appear at the surface itself they are often but thinly covered by poor stony soils, whilst the many steep slopes as well as the broken character of the relief may make farming difficult. On the whole man has sought out for himself the more sheltered situations for his farms, his villages and his towns. They nestle in valleys; they flourish and spread only where the larger tracts of flat land or more fertile soil occur or where man has been attracted to otherwise inhospitable surroundings by stores of mineral wealth. The higher, poorer, wetter or less accessible parts of Highland Britain have been left largely to nature. There are vast stretches of moorland, some of it covering land from which the original forest cover has been removed, as well as mosses and bogs, scrub woodland and forests. We may summarise the position by saying that in Highland Britain human settlement is essentially discontinuous: the cultivated areas occupy valleys and plains separated by large expanses of uncultivated hill lands. This is well shown in the aerial view of a Lakeland valley in Plate I.

Lowland Britain offers a striking contrast in many ways. Though so much less rugged, there are few parts where level land is uninterrupted by hills and such true plains as do exist are to a considerable degree the result of man's handiwork—as the large stretch of the drained fenlands of eastern England bears witness. Lowland Britain

FIG 1.—Highland Britain and Lowland Britain
The parts shown in black are built up of rocks geologically older than these which carry our coal seams

is best described as an undulating lowland where lines of low hills are separated by broad open valleys and where " islands " of upland break the monotony of the more level areas. Even the highest of the hills scarcely ever exceed a thousand feet above sea level, though many of the ridges reach 600 or 700 feet. The environment is kinder ; the soils tend to be deeper and richer, there are few steep slopes to interrupt cultivation and plough lands are to be found right to the tops of the hills. There is little to hinder man's use of the whole: human settlement is essentially continuous and the cultivated land of one parish merges into that of the next. Villages and towns are closely and evenly scattered; their siting has sometimes been dictated by convenience of a water supply, sometimes by situation on a natural routeway, sometimes just to maintain an even spacing of settlements. It follows that the greater part of Lowland Britain is occupied by farmland—by "cultivated" or "improved" land, which includes both plough and grass land—and that such moorlands, heaths, "wastes" and other unimproved lands as occur, do so as islands interrupting the otherwise continuous farmland and coinciding with patches of poorer soils.

It so happens that the rain-bearing winds in Britain blow from the west, especially from the south-west, and that the hill masses of Britain are in the west of the country. As a result Highland Britain as a whole gets a heavy rainfall, for the hills stand in the path of the rain-bearing winds, whilst most of the Lowland Britain is relatively dry. Taking a very general figure, most of Highland Britain gets more than 30 inches of rain a year while most of Lowland Britain gets less. Since the trouble with our climate is that we tend to get too much rain and too little sun it is clear that differences in climate reinforce those due to other features between Highland and Lowland Britain. In the far north-west, for example, where lowland does exist it is of comparatively little value because heavy rainfall and constant cloud result in waterlogged conditions and almost useless bogs occur where otherwise good farmland might be found. The contrast in rainfall is shown in the accompanying map, while the effect of the combination of high ground and high rainfall is well brought out in Fig. 3 which shows the general distribution of moorland.

Although the chief visible contrasts between Highland and Lowland Britain are thus to be found in elevation, relief of the ground and soils, these are, in fact, the results of the underlying geological structure. The rocks which underly the hills are for the most part old even in the geological sense and in the course of ages have tended to become indurated and hard or at least resistant to weathering and hence tend to form hill masses rising above the general level. The rocks which underly the lowlands on the other hand are younger in the geological sense, though often very ancient if measured in years. They are softer or less resistant to weathering—there is evidence of this in the muddy streams to be seen after heavy rain, showing that soil or mud or silt is being swept off the land and carried seawards.

It happens that Highland Britain is made up mainly of rocks which are older than the Coal Measures (which contain the bulk of workable coal seams in this country) whereas Lowland Britain is made up mainly of rocks which are younger than the Coal Measures (see page 17). It is for this reason that we find most of the coalfields of Britain on the

FIG. 2.—The Mean Annual Rainfall of the British Isles, showing the contrast between the wetter west and the drier east

margins of the highlands and the lowlands. The greatest of them all, the Yorkshire, Nottinghamshire and Derbyshire Coalfield, stretches up on to the Pennines and down on to the lowlands ; so do the coalfields of Lancashire and of Northumberland and Durham. The great South Wales field lies just north of the fertile Vale of Glamorgan and amongst the moorland heights.

As a prelude to understanding the different regions of Britain and

FIG. 3.—The Chief Moorland Areas of Great Britain, showing the close correspondence with areas of heavy rainfall (Fig. 2) and their association with Highland Britain

consequently the varied habitats in which our plants and animals live it is thus essential to understand the geology and structure of the ground and to appreciate something of the long and complicated geological history of the country.

READING THE ROCKS

IT HAS long been known that the solid rocks which build up the earth's crust sometimes contain remains of animals and plants. A slab of shale from the tip-heap of a coal mine may be split open to reveal a beautifully preserved and delicate leaf which, whilst bearing a superficial resemblance to some ferns, on closer examination is found to be different from anything now living. In one of the literary classics which geology has given the English language the young Scottish stone-mason, Hugh Miller, has described the thrill of the chase when his hammer was used to make the Old Red Sandstone rocks give up entombed fragments—so clearly of fish yet so utterly different from their counterparts of the present day. Sometimes it is merely the footprint of some long-extinct reptile, sometimes the actual bones or teeth ; at other times it may be the remains of some minute creatures only revealed by the microscope which excite interest and inquiry into the origin of fossils, as all these remains of the animals and plants of the past are called.

Fossils were once hailed as incontrovertible evidence of the reality of Noah's flood. Since it was soon made clear that they were to be found in different rocks and at different levels there arose the idea of several successive creations each in its turn overwhelmed by a great deluge. Forming as it were a mantle over many parts of the country are superficial deposits of clay, sand and gravel, to which reference will be made later. These " drift " deposits not infrequently contain shells and other fossils and the inference that these deposits were laid down by the latest flood was so obvious that they were called by the eighteenth century geologists " diluvium " (Latin, *diluvium*, a deluge or flood) or " diluvial deposits."

A great advance was made when William Smith (1769-1839), who has very rightly been called the Father of English Geology, showed that the same fossils (i.e. different specimens of the same species) were to be found in different parts of the country, sometimes in the same type of rock but sometimes in rocks of different types. If several different species were associated together in one area then the same ones would be associated together in another. So he introduced the *Law of Strata identified by fossils,* and was able to produce the first geological map of England and Wales (published 1815). Two limestones from different parts of the country might appear to be very similar but if they contained different sets of fossils the inference was that they were of different ages. If on the other hand a sandstone from one region contained fossils identical with those from a shale in another the inference was that the two rocks were being formed at the same time—that they were of the same geological age or " synchronous " and could be " correlated." It was found that when a species died out or " became extinct " it did not reappear in later rocks. Of course this was quite consistent with the idea of a succession of separate creations and it was not till much later that the theory of evolution permitted the tracing of the relationship between the fossils of one set of rocks and those of another.

The determination of the geological age of rocks by the fossils they contain is one of the two fundamental principles underlying the whole study of historical geology. The other is the *Law of Superposition.* Where one bed of rock rests upon another it is presumed that the upper bed was laid down after the lower and hence that the upper bed is the younger. A large proportion of the fossil-bearing rocks are sedimentary rocks (i.e. they were laid down as sediments under water—in the sea or in freshwater)—and this law is true for nearly all such rocks. It is also true for streams of lava poured out from volcanoes or associated beds of ashes. Pompeii was there before the ashes by which the city was buried. But the Law of Superposition only remains true so long as the original order of the rocks has remained unchanged. With earthquakes and mountain-building movements the original order may be changed—the rocks may be folded, or even bent right over so that the original order is reversed. But reversal of the order in this way is the exception, not the rule, and can be detected by detailed survey. No one who has spent a holiday on the magnificent coast of north Cornwall can fail to have noticed how folded and broken are the rocks exposed along the sea cliffs. Examples are shown in Plates III and IV.

Presuming the original order to have been maintained, if the upper bed in one locality be traced laterally it may be found to pass under still higher beds in another so that the higher beds in the latter area are still newer. In this way a whole succession of strata may be built up—from the very oldest at the bottom to those still being formed at the top. Such a succession has, in fact, been constructed as a result of patient research and forms what is sometimes called the geological column. It must be realised that the rocks of the geological column are not to be found complete in any one area.

It must not be thought that if a hole were bored in the earth's surface it would pass through all the rocks shown in the geological column. At the present day deposits of sand, silt and mud are being formed in the shallow waters near the estuaries or deltas of the great rivers of the world while other deposits, some of them consisting mainly of the hard parts of organisms living in the water, are being formed over the floors of most of the seas and oceans. In the lakes of the world other deposits are being laid down and even on parts of the land surface deep layers of sand and dust brought by wind are being spread over the older rocks. This is quite clearly seen where, as in the Culbin Sands of Morayshire shown on Plate XII, sand dunes are burying growing vegetation. These are all areas where *deposition* is taking place and where new strata are being formed. But over most of the land the rocks are being worn away by the combined action of rain, wind, sun, frost, running water and moving ice and its surface slowly but inevitably lowered. These are areas of *denudation* (Latin, *denudo*, I lay bare) and to them may be added the margins of the oceans where waves beat against the shores and wear them away. There will be no deposits in such areas to mark the present day and it was the same in the past. Thus beds present in one locality may be absent in another so that in the latter place there is a gap in the succession. Such a gap may indicate that the region concerned formed part of a land mass at the time in question or came otherwise under the influence of denudation. When the region again sank below sea-level and strata were again deposited it was perhaps after a lapse of many millions of years. Here is a "stratigraphical break" between the older and younger rocks. The younger rocks are said to rest "unconformably" on the older in those cases where the older had been folded and denuded in the meantime. A typical unconformity is shown in Plate IVB.

There is another difficulty in reconstructing the stratigraphical column. When one bed of rock or stratum is traced laterally it may change its character and unless the whole change can be traced and a limestone, for example, found to pass laterally into a shale or sandstone, it may be difficult to say that the limestone in the one place *is* of the same age as the sandstone in another. Of course if both types of rock

L. D. STAMP

Culm Measure sandstones at Bude, Cornwall (See page 52)

L. D. STAMP

Vertical sea-cliffs at Crackington Haven, Cornwall

The plane of attack by wave action is at the base of the cliffs between tide marks and is indicated in the upper picture by the rounded boulders and stones (See page 52)

L. D. STAMP

Cotton-grass moor near West Country Inn, Hartland, Devon (Page 215)
The peneplane surface is ill-drained: a slight rise permits better drainage and cultivation

L. D. STAMP

Vertical strata, Compass Point, Bude, Cornwall
The form of the ground has no relation to the underlying strata (See page 52 and page 148)

contain the same fossils the answer is easy, but just as different habitats at the present day—muddy waters and clear lime-rich waters—may not have even a single species in common, so it happened in the past, and the fauna of a limestone may be completely different from the fauna of a bed of shale of the same age. To take a specific example, the beds known collectively as the Old Red Sandstone were laid down in freshwater lakes at the same time as the marine beds of the Devonian were being deposited elsewhere. In such cases the rare instances where the faunas are mixed, or there are "marine bands" representing incursions of the sea in the midst of a fresh-water succession, are invaluable in establishing the essential correlation. Thus the evidence which the geologist has to piece together is at the best fragmentary: it is rarely too that the rocks he wishes to study are "exposed" over large areas. In a country such as Britain the surface is hidden by soil and vegetation and only in some of the higher mountainous areas or along sea cliffs do the bare rocks outcrop at the surface. Elsewhere the geologist has to seek his evidence in quarries, mines, railway cuttings, well-borings, casual excavations for drains and sewers and even in some cases may be faced with the necessity of opening up a special pit in a crucial spot.

The rocks which are seen in the Stratigraphical Column were deposited over an immense period of time. Time is continuous, but there are certain natural phenomena which serve to divide it into definite units. The phenomenon of day and night serves to define one unit of time—the day; the movement of the earth on its orbit round the sun definites another—the year. Larger units than the year are difficult to define but just as the astronomer uses a "light-year"—to define an enormous *distance*, so the geologist needs a larger unit than the year. The historian frequently takes the time between two important events to define a period; thus when we talk of Tudor Times we mean the period when the Tudor kings were on the English throne, though we are able to define this period accurately in years—from the accession of Henry VII in 1485 to the death of Queen Elizabeth in 1603. The prehistorian is no longer able to measure his periods so accurately: he is obliged to define them in terms of the works of man in the periods concerned. The geologist, in his turn, has to deal with the vast periods of time which elapsed before the appearance of man on the surface of the earth; for the definition or delimitation of such periods the year is an inadequate unit. No one would hand a traveller going on a long sea voyage a six-inch ruler and ask him to measure thereby the distances between the ports en route. Yet the voyager, by careful observation of time and direction, might be able to give a very fair account of the relative positions of the points touched, a close estimate of the distances between them and a good general account of their chief features. It would depend on his power of accurate observations and of using all the available evidence in its appropriate place. Thus the geologist has built up a good general picture of the evolution of the earth's surface, a picture which is continually gaining in accuracy, and the geological time-scale is divided into a few great eras and a number of periods. The smallest unit of geological time is the *hemera*, usually named after a dominant animal or plant which was living at the time. A difficulty is that the animal or plant *may* have been local in its distribution, so that its absence from the sequence in a given locality is scarcely sufficient evidence that no deposition of beds was going on there at that time. A somewhat larger unit is the *zone* which, though named after a characteristic fossil, is usually to be defined by a characteristic associated series of fossils. A number of zones normally comprise a *formation* of rocks, while several formations make up a *system* of rocks. Thus we talk about the Chalk and the Lower Greensand as two of the formations in the Cretaceous System of rocks. But the word system refers to the *rocks* in the geological column: the rocks in a system were laid down in the period of time known as a geological *period* so that the measure of time concerned in this case is the Cretaceous Period. A number of periods are included in each of the four great *eras* into which geological time has been divided since the general appearance of life on the surface of the globe—i.e. since the deposition of the rocks which contain the earliest recognisable fossils. Even before that the earth had a long and complicated history which is gradually being unravelled and lowly forms of life doubtless existed but have left little or no trace.

On the general divisions of the geological time-scale and on the sequence of the periods all geologists are agreed though there is constant discussion regarding the exact definitions of the periods and whether a given bed of rock was laid down at the end of one period or the beginning of the next.

The layman is constantly demanding to be told the age of a given bed in years and amongst geologists themselves the age of the earth has always been a fascinating subject. One ingenious calculation

Era	Period or System	Approximate Duration in Years	Cycles of Earth Movement
QUATERNARY	RECENT (HOLOCENE) PLEISTOCENE		
TERTIARY OR CENOZOIC Often written Cainozoic or Kainozoic	PLIOCENE	65,000,000	
	MIOCENE		Alpine
	OLIGOCENE		
	EOCENE		
SECONDARY OR MESOZOIC.	CRETACEOUS	55,000,000	
	JURASSIC	30,000,000	
	RHAETIC	10,000,000	
	TRIASSIC	30,000,000	
PRIMARY OR PALÆOZOIC	PERMIAN	30,000,000	
	CARBONIFEROUS	60,000,000	Armorican
	DEVONIAN	40,000,000	
	SILURIAN	30,000,000	Caledonian
	ORDOVICIAN	50,000,000	
	CAMBRIAN	100,000,000	
PRE-CAMBRIAN OR EOZOIC	PRE-CAMBRIAN		Charnian

FIG. 4.—The Geological Column with the names of the geological periods and an approximate time scale in years. See also p. 17

worked out the amount of dissolved salts carried down to the sea every year by the rivers of the world and consequently how long it would have taken the ocean, presuming the water of the ocean to have been fresh originally, to have reached its present degree of salinity. A rough method at best, it breaks down as there is no evidence that the waters of the world ocean were originally fresh. In recent years a method of estimating geological time in years has been devised and used with considerable success. There are certain elements—the radioactive elements—which undergo disintegration at a constant but very slow rate which can be and has been measured. When a minute crystal of a radioactive mineral is enclosed in a larger crystal of certain other minerals—such as the dark mica, biotite—the emanations from the radioactive mineral cause a visible change in the surrounding mineral. When studied in section under the microscope the size of the zone of alteration, or "pleochroic halo," affords a means of measurement of the time which has elapsed since the original formation of the rock. Another method is by the very accurate chemical analysis of the unaltered radioactive substance proportionate to the amount of the final end-products of its disintegration.

Piecing together the evidence, the geological column and the approximate duration of each period are in Fig. 4.

The names of the periods are reminders both of the richness of the British Isles in its varied geology where all the great systems are represented and also of the pioneer part played by British scientists in the geological field. The Cambrian takes its name from Cambria or Wales; the Ordovician and Silurian from two tribes of ancient Britons who lived on the Welsh borderland where these rocks are well developed and where they were first described. The name Devonian is from the county of Devon. Permian is a name which honours the pioneer studies of the British geologist Murchison in the province of Perm at the request of the Russians. Carboniferous (carbon- or coal-bearing) and Cretaceous (chalky) are descriptive of British conditions whilst the names of the divisions of the Tertiary are reminders of the Greek scholarship of Charles Lyell and his followers. The Jurassic (Jura Mountains) combines the older English Lias and Oolites; the Trias takes its name from the three-fold division typical of that system in Germany. Only the Rhaetic (Rhaetic Alps) is poorly represented in Britain. The Quaternary is not really comparable in duration or importance with the other great eras.

Era	Period or System	Approximate Duration in Million Years
QUATERNARY	Recent or Holocene	65
	Pleistocene	
TERTIARY OR CENOZOIC*	Pliocene	
	Miocene	
	Oligocene	
	Eocene	
SECONDARY OR MESOZOIC	Cretaceous	55
	Jurassic	30
	Rhaetic	10
	Triassic	30
PRIMARY OR PALAEOZOIC	Permian	30
	Carboniferous	60
	Devonian	40
	Silurian	30
	Ordovician	50
	Cambrian	100
PRE-CAMBRIAN OR EOZOIC		

The coal-bearing strata of the Coal Measures form the upper division of the Carboniferous.

Recent studies and more accurate measurements have made it necessary to revise the figures given on Fig. 4 and in the table above which shows the earliest Cambrian rocks to have been deposited about 500 million years ago. A more likely figure is 650 million years. Among the pre-Cambrian metamorphic rocks of Britain are some 2,500 million years old; in other parts of the world are rocks believed to have been formed 6,500 million years ago.

* Often written Cainozoic or Kainozoic.

CHAPTER 3

EARTH HISTORY—TIME AND LIFE

WE ARE fortunately living in one of the quiet periods of the earth's history, though not at a time when it has been quiet for so long that the surface has been worn down to a dull and monotonous level. Although the accurate instrumental observation of earthquake phenomena demonstrates that the earth's crust is rarely absolutely still, the occasional earthquakes which do occur, however severe they may be, are in reality but very gentle reminders of those periods in the history of the earth when the whole surface must have been torn by the most violent and long-continued cataclysms which bent and folded and broke the hardest rocks ; which caused whole blocks of the earth's crust to be thrust many scores of miles over other blocks ; which caused continents to sink below the waves of the sea, or which caused vast tracts of the ocean bed to be raised to form the world's highest mountains. The earthquakes of to-day are like the final murmurs of a great storm which has passed. Even they tend to occur in certain defined earthquake belts and are reminders that the earth's crust has certain lines and zones of weakness whereas other parts are relatively stable.

If we exclude the great mountain-building movements which took place in the dim early days of the earth's history—in the Pre-Cambrian —there have been three great periods of mountain building or " orogenesis " (Greek, *oros* mountain ; *genesis* origin, creation) as far as Europe is concerned. Each of these has played a major part in determining the present-day features of Britain. These three great mountain-building periods were :

(*a*) at the end of the Silurian and beginning of the Devonian periods—the Caledonian earth-movements, so called because they built up the great mountains which have since been worn down to form the Highlands of Scotland (Caledonia) ;

(*b*) at the end of the Carboniferous and beginning of the Permian periods—the Armorican or Hercynian earth-movements, so called because they caused the great folding of the rocks seen in Brittany (Armorica) and the Hartz Mountains of Germany. In Britain they caused the uplift of the Pennines, the Malvern and Mendip Hills and folded the Coal Measures into basins.

(*c*) in the middle of the Tertiary era—the Alpine earth-movements, so called because they were responsible for the rise of the Alps as well as of many of the great mountain chains of the world to-day but which affected Britain much less than the previous movements.

Of the still earlier earth-movements at least one left very important marks in Britain—the Lewisian, which caused the folds in the rocks in the extreme north-west of Scotland and which may have been contemporary with the folding of the ancient rocks which peep from beneath a cover of later strata in the Charnwood Forest of Leicestershire.

In each case the mountain-building movements gathered strength slowly as with a developing storm and gradually reached a peak when the whole earth must have experienced a constant succession of gigantic earthquakes. Then gradually they must have died away again, the whole cycle stretching over an immense period of time. The result of these earth-building or orogenic movements was to form a series of gigantic wrinkles in the crust of the earth—these are the main mountain chains—between which are broad areas but little disturbed—the " tectonic " basins (Greek : *tektonikos*, related to building—i.e. not formed by later excavation). Sometimes these basins were below sea-level and became the areas of sedimentation in the succeeding periods, while the surrounding mountains as soon as formed were attacked by the forces of denudation which started to wear them down. So we get the idea of the geological history of the earth moving in great cycles. The first is what may be called the major cycle of denudation. This may be considered to begin when earth movements have caused land to rise above the level of the waters in the surrounding ocean. No sooner does this happen than the forces of sub-aerial denudation get to work. The heat of the sun heats the rocks and the different minerals of which they are composed have differential rates of expansion so that, especially with nightly cooling, the rocks are disrupted and a peeling or exfoliation (Latin : *folium*, a leaf) by successive layers takes place. This is sometimes called onion weathering and is well seen in hot dry countries at the present day. The direct action of the sun is called insolation.

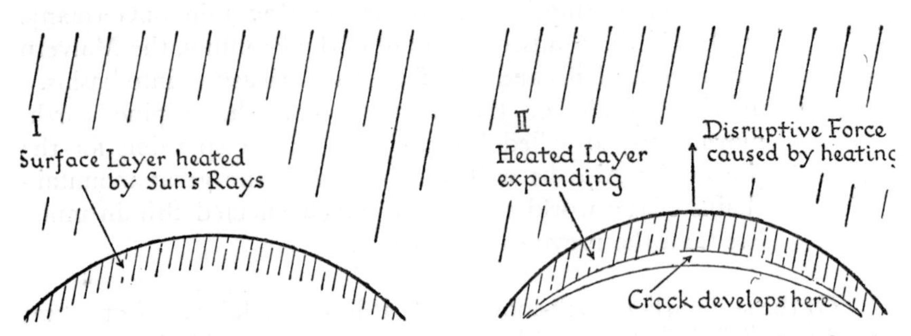

FIG. 5.—Diagrams showing the mechanism of *exfoliation* or onion weathering of rocks under the sun's heat

Falling rain has a direct mechanical effect in washing away the finer particles, a less direct effect by dissolving some of the less stable minerals and an indirect effect by soaking into crevices. There it may be frozen and the water in changing to ice expands so that the crack is widened. This is the basis of frost action, through which great blocks may be split off from mountains and fall to lower levels as screes. Wind, too, plays its part by blowing away the finer dust and sand whilst strong wind armed with sharp sand particles is a powerful abrading agent. In newly formed mountain areas gravity itself plays a large part—for example in the formation of screes. Both in mountain areas and at lower levels landslides are by no means unknown. Gravity also causes the well-known phenomenon of soil creep, whereby soil gradually slides downhill. The process is seen at work in Plate 9B. Rain collects together to form mountain torrents which in turn unite to form swift rivers sweeping masses of debris always from higher to lower levels, from the land towards the sea. The eroding and transporting action of running water is paralleled in colder climates by the action of moving ice—glaciers which move slowly but inexorably down valleys or great ice sheets which ride over the whole surface of the land, scooping out hollows where the rocks are soft, smoothing and polishing them where they are hard. In tundra lands the sub-soil remains permanently frozen whilst the surface thaws in summer and, where there are steep slopes, masses of sludge slide downhill, the whole process being called solifluction. On the margins of the seas and oceans wave action is a powerful force in wearing away the newly formed lands.

Whilst the major surface features of Britain owe their origin to the mountain-building movements of the past and to the character of the rocks which make up the land masses, many of the most striking scenic details are the result essentially of the different processes of weathering on varied rocks. In high mountain areas frost plays a large part and accounts for the angular rock surfaces such as those seen on Striding Edge (Plate XVIB) or in Snowdonia (Plate 8A) or on Cader Idris (Plate XXIX). Sometimes the sculpturing action of frost produces fantastic results, as in the well-known Sphinx Rock on Great Gable in Lakeland. Screes of fallen angular blocks and fragments of rock, most of them broken off by frost action, are a well-known feature in all mountain areas and sometimes dominate the landscape. Plate 30B shows the famous screes on the south side of Wastwater. Blocks of rock dislodged by the undercutting action of the sea and the action of rain form screes along many sea cliffs ; a typical example from Cornwall has been shown in Plate 8B to illustrate the angle of rest assumed by loose rock of average character. The angle is much lower where rocks such as clay-shales become slippery when wet, and is lowest where the actual rock may " flow " when wet, which is the case with clay.

Onion weathering under the influence of the sun leaves hard, rounded cores of rock. In tropical countries, these may be almost true spheres ; in this country such " cores " scattered over the country are familiar in many granite areas. A good example may be seen on Crousa Common (The Lizard, Cornwall), whilst the interesting weathering of granite, seen in such " tors " as those of Dartmoor (Plate XXVII) is to be ascribed mainly to the same action.

The most interesting results are seen where the original rock varies in hardness. A sandstone, for example, may be indurated along certain lines and the denuding agent whether wind, rain, running water or the sea finds out the pockets of softer sand and scoops them out. The interestingly fretted rock shown in Plate VI is actually the result of the action of the sea, but a very similar appearance might be due to wind action. Where a rock is fractured rain washes out the loose, crushed rock and produces striking cliffs such as those shown in Plate 1B. Even in Lowland Britain the " High Rocks " of Tunbridge Wells are simple examples of differential weathering.

Immediately after a great earth-building movement the deposits which fill the hollows—the tectonic valleys and basins—are coarse and often consist of angular blocks which are actually screes and may become consolidated to form a "breccia." Beds of roughly rounded boulders and large pebbles may be deposited by swift streams to become consolidated later as conglomerates and pebble beds. Plate VIIIB shows an example from the Lake District of such boulders being swept down by a stream in flood. As time goes on the mountains are worn down, yield less material and the beds laid down in the basins and seas become finer grained in character—sands and silts and muds, which may become consolidated respectively into sandstones, siltstones and shales. In the later stages of the cycle muds and clays will definitely predominate and when the lands have been worn down almost to plains (called "peneplanes" or "peneplains"—Latin: *pene*, almost) they will yield so little sediment that the waters of the surrounding seas may become quite clear. These conditions of clear tranquil water are those under which corals flourish and also other organisms which build up their hard parts of calcium carbonate; thus the deposits then formed are often limestones. The cycle of denudation on the land and of sedimentation in the water is brought to a close by earth movements, it may be slight at first, which herald the oncoming of a new storm. More often the major cycle of events is varied by minor earth movements—it may be the so-called "eustatic" movements, not of folding of the earth's crust, but of the gentle elevation or depression of blocks of it relative to the level of the waters—so that minor cycles of sedimentation occur within the major. This is well illustrated in the geographical evolution of the British Isles (see page 142).

So far nothing has been said regarding what is now known of the structure of the earth as a whole. It cannot be too forcibly stated that the old concept of a solid crust, rather like the skin of an apple, covering a molten interior, is entirely wrong and that the simple deduction that the whole was cooling and contracting so that wrinkles—which were the mountain ranges—were being formed just as when an apple dries is equally false. We now know that there is a central sphere, solid and very heavy and probably consisting of an alloy of iron and nickel—thus agreeing in composition with some of the meteorites which from time to time fall on the earth's surface. This iron-nickel core accounts for the magnetic phenomena of the earth. Enveloping this is the crust, in all about 700 miles thick—a figure which may be compared with a height of 5 miles for the highest mountain and a depth of 6 miles for the deepest ocean. It is well known that there is a rapid increase

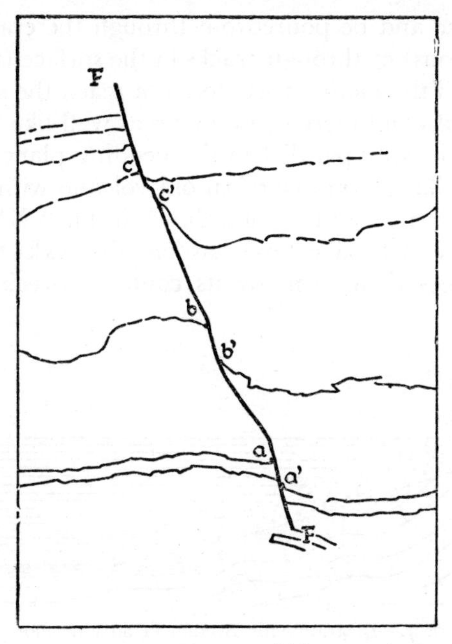

Fig. 6.—Diagram of the Fault shown in Plate 1A. This is a typical example of a very small normal fault. The fault plane separates the downthrow side on the right from the upthrow side on the left. The angle which the fault plane makes with the vertical is the hade ; the vertical displacement (here only a few inches, though in big faults it may be thousands of feet) is the throw. Normal faults occur under tension whereas thrust faults and structures such as are shown in Fig. 72 occur under extreme compression.

in temperature as one goes downwards in the crust so that even in a deep mine it is almost unbearably hot. It does not necessarily follow that the solid core of the earth is extremely hot, since it is now known that heat accumulates in the lower layers of the crust through radio-activity. What is important is not the temperature of the central core but of the crust. At no great depth the temperature must be such that all rocks would be molten were they not kept in a solid or more probably a plastic condition by the pressure of the solid rocks above. Towards the end of a major cycle of denudation, however, so much material has been removed from one part of the surface of the crust to another that the pressure is lessened over the land. Some of the underlying heated layer becomes actually molten and seeks to find weak spots or lines in the crust through which it can escape. It may reach the surface and be poured out through the craters of volcanoes (volcanic eruptions) or through cracks in the surface (fissure eruptions) as lava. Some of the molten rock does not reach the surface but forces its way into cracks and there consolidates as wall-like masses or dykes ; or it may force its way parallel to the bedding planes of sediments to form sills. A striking example of an old volcano with associated sill is found in Arthur's Seat, Edinburgh, shown in Plate XVIA. In all these cases the molten rock bakes and hardens the rocks through which it passes—it changes their form by its contact (Greek : *meta-* change,

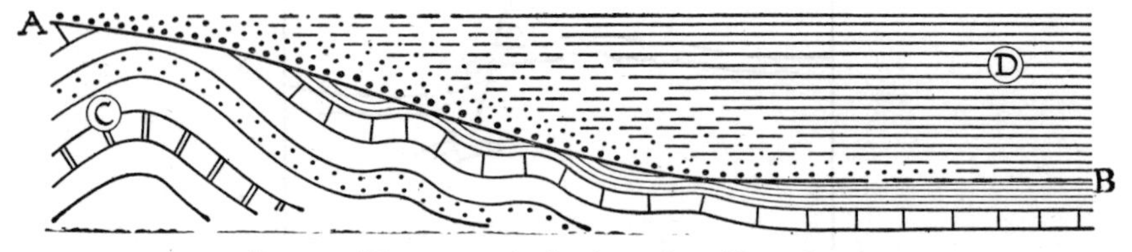

FIG. 7.—Diagrammatic Section of an Unconformity

A—B is the plane of the unconformity. After the deposition of the group of beds marked C they were gently folded by earth-building movements and were subjected to denudation. Gentle subsidence followed so that the group of beds marked D were deposited gradually over a larger and larger area—they rest unconformably on the older series and at the same time overstep them. In the centre of the basin fine-grained shales were deposited and the diagram suggests that sedimentation was almost continuous. Towards the margins of the basin the fine-grained deposits pass laterally into sands and other coarser sediments and to shore deposits.

This section represents diagrammatically the relationship between the Silurian and the underlying Ordovician in the Welsh Borderlands, described on p. 110. See also Plate IV B.

morphe form, hence the process is called contact metamorphism). Thus clay and shale are baked into hard slatey rocks, limestone is changed into marble. Some of the molten rock is very fluid when it is first poured out and spreads evenly over a wide surface as did the basalts of Northern Ireland ; sometimes it formed hexagonal columns on cooling as at Giant's Causeway and the Island of Staffa (see Plate XVA). In other cases the molten rock was very sticky and consolidated almost on the spot—the famous conical " spire " of Mont Pelé of Martinique in the West Indies is the best modern example of this (it was formed during the disastrous eruption in 1902) but there are many examples from earlier periods in the British Isles, such as Ailsa Craig in the Firth of Clyde. Even more important, though it cannot be observed at the surface, is the underground movement of great masses of molten "magma." At the height of great earth-building movements the magma is squeezed into the core of mountain ranges so that millions of years afterwards, when the mountains have been denuded down to their roots, this core is exposed. A typical rock so formed is granite. The metamorphism caused by a huge granite mass taking eons to cool can perhaps be imagined rather than described and the "metamorphic aureole" is often very extensive; it is economically important because of the valuable metallic minerals which are associated with the gases and heated liquids given off by the magma. These latter often find their way into cracks or veins and there the minerals are deposited—hence the association of ores of tin and copper with the metamorphic aureoles of the granites of Devon and Cornwall. These Devon and Cornish granite masses are the roots or cores of giant mountains formed by the Armorican earth movements but long since worn down almost to a level surface. It is clear that there is a definite cycle of igneous activity associated with a cycle of earth movements—volcanic activity heralding the oncoming storm; intrusion and movement of huge underground masses at the height of the storm; and finally renewed volcanic activity when the storm is dying away. The few volcanoes on the surface to-day which are active may be regarded as the last remnants of the once wide-spread activity at the end of the Alpine earth movements. Many of those are at the end of their lives—dormant or even extinct or merely giving off vapours ("solfataric stage"). Hot springs and geysers are indications of a nearly dead volcanic area.

One of the still unsolved puzzles of earth history is whether or not there have been true climatic cycles in the past. There is no doubt that at several periods there have been ice ages, though perhaps nothing as severe as that which overwhelmed the northern hemisphere so recently in geological time that man was already well established and hence known simply as "the Ice Age." We know definitely that some of the red rocks found in the British Isles—such as the Permian or New Red Sandstone—were laid down under desert conditions and there must have been other times when what is now our country must have been wet and hot. Many of the phenomena, however, may be

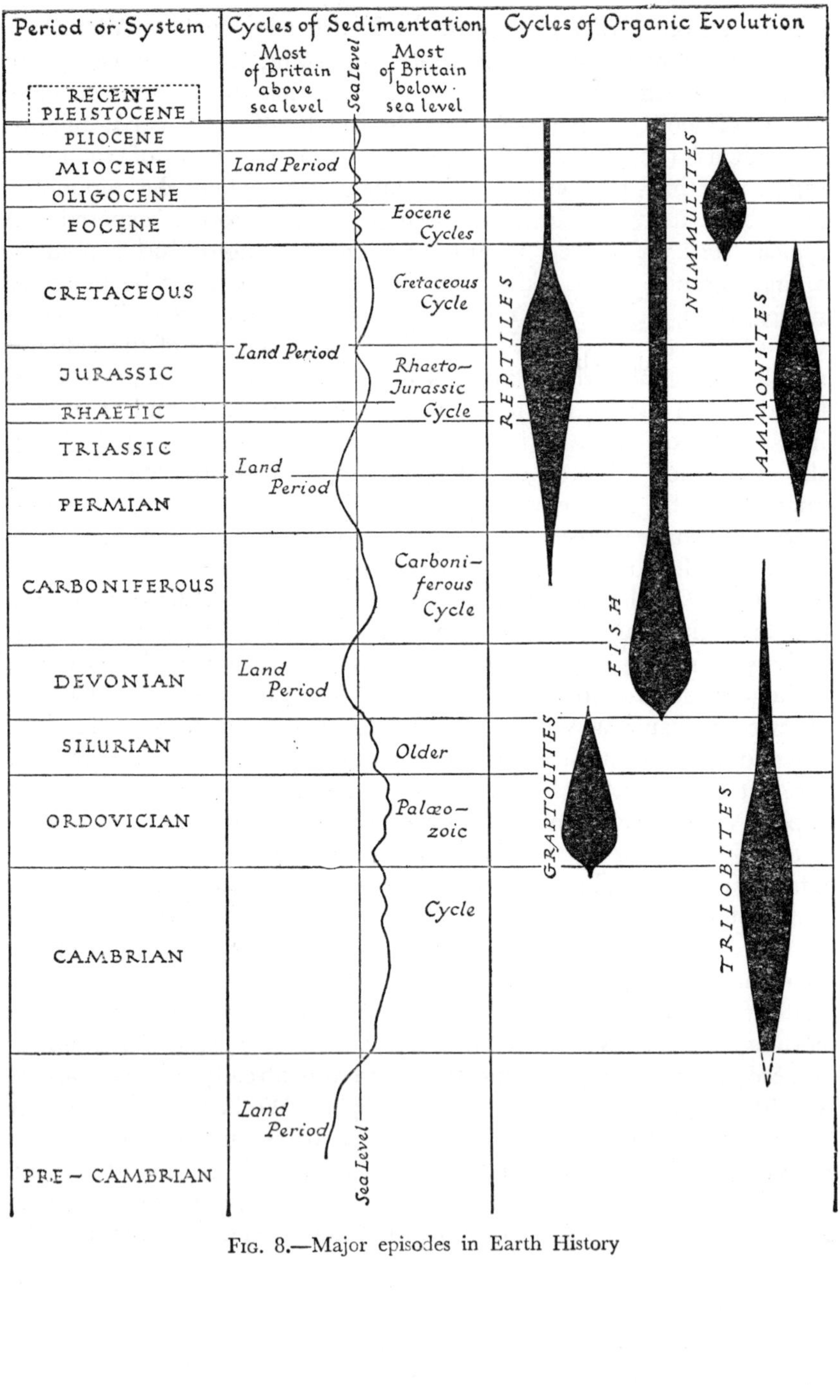

Period or System	Cycles of Sedimentation: Most of Britain above sea level	Cycles of Sedimentation: Most of Britain below sea level
RECENT / PLEISTOCENE		
PLIOCENE		
MIOCENE	Land Period	
OLIGOCENE		
EOCENE		Eocene Cycles
CRETACEOUS		Cretaceous Cycle
	Land Period	
JURASSIC		Rhaeto-Jurassic
RHAETIC		Cycle
TRIASSIC	Land	
PERMIAN	Period	
CARBONIFEROUS		Carboniferous Cycle
DEVONIAN	Land Period	
SILURIAN		Older
ORDOVICIAN		Palæozoic
CAMBRIAN		Cycle
PRE-CAMBRIAN	Land Period	

FIG. 8.—Major episodes in Earth History

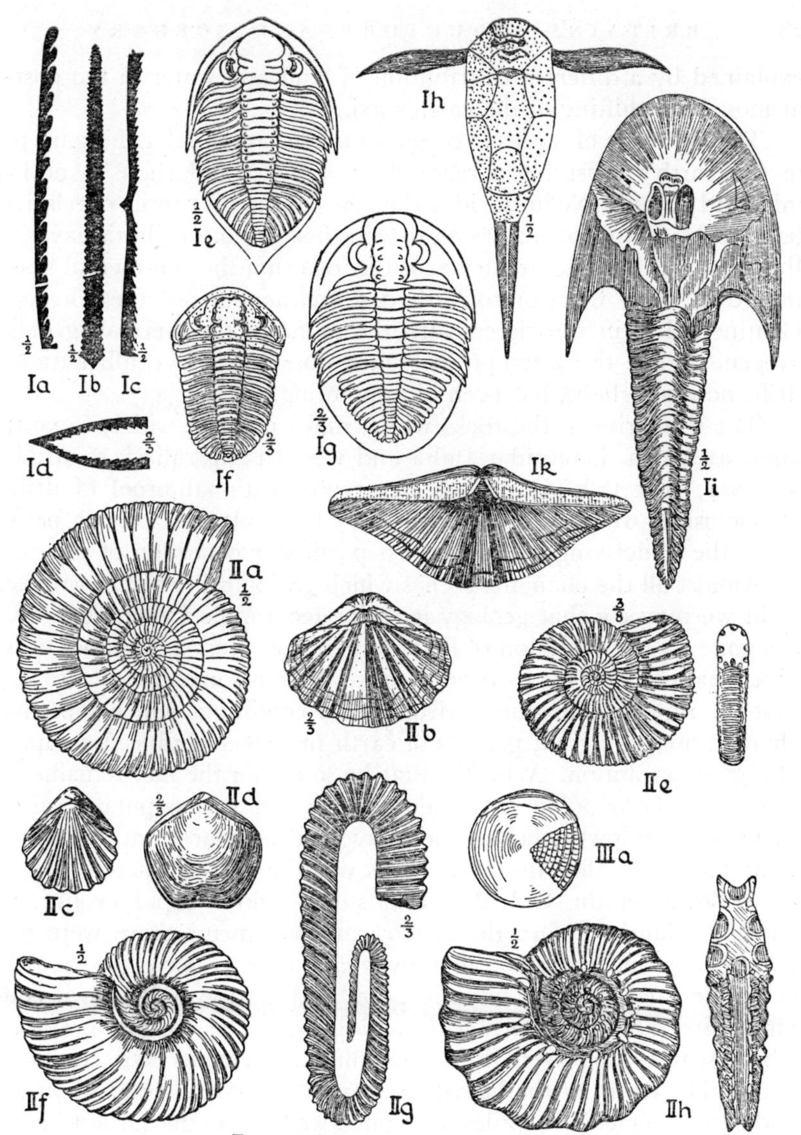

FIG. 9.—Some characteristic Fossils

Those numbered I are Palaeozoic ; those numbered II are Mesozoic and III Cenozoic.

Graptolites :—Ia *Monograptus* (Silurian) ; Ib (*Diplograptus* (Ordovician) ; Ic *Didymograptus extensus* (Ordovician) ; Id *D. murchisoni* (Ordovician).

Trilobites :—Ie *Phacops caudatus* (Silurian) ; If *Calymene blumenbachi* (Silurian) ; Ig *Ogygia buchi* (Ordovician).

Primitive fish :—Ih *Pterichthys milleri* and Ii *Cephalaspis lyelli* (Old Red Sandstone).

Brachiopods :–Ik *Spirifer verneuilli* (Devonian) ; IIb *Spiriferina walcotti* ; IIc *Rhynchonella rimosa* and IId *Waldheimia numismalis* (all Lias).

Ammonites :—IIa *Dactylioceras commune* (Lias) ; IIe *Perisphinctes biplex* (Jurassic, Kimeridgian) ; IIf *Hoplites splendens* (Cretaceous : Gault) ; IIg *Hamites attenuatus* (Cretaceous : Gault) ; IIh *Hoplites auritus* (Cretaceous : Greensand).

Nummulites :—IIIa *Nummulites laevigatus* (Eocene).

explained by a different distribution of land and water in the past or at most by a shifting of the earth's axis.

The existence of a plastic or semi-molten layer under the solid part of the earth's crust has already been argued and there is nothing inherently impossible in the idea that the continents consist of relatively light rocks and form masses as it were floating on a plastic layer. If this is so, it is but one step towards the idea that the continental masses may drift away from or towards one another—hence the Theory of Continental Drift associated with the name of the German geologist Wegener. But the attempt to secure observational confirmation of drift, however slight, has been disappointing.

The similarity in the rocks which make up such widely separated lands as Africa, Peninsular India and western Australia is so striking, however, that this absence of direct observational proof of drift is inconclusive. It may be that drift only takes place at certain periods when the underlying rocks are in a particular condition of plasticity.

Amidst all the changing scenes which geological time has witnessed—for we may say that geology is really geographical evolution—there has gone on the evolution of living organisms. Just as in times of war things happen and life is speeded up so there is some evidence to show that the rapidly changing environmental conditions which must have characterised the great periods of earth movement induced a rapidity of organic evolution. Whether that be so or not the fact remains that before the Caledonian movements the world was populated almost exclusively by lowly plants which have left few traces and by invertebrate animals. After the movements was the age of fishes—the many weird forms of the Old Red Sandstone—and a rapid evolution of fern-like plants. Before the Armorican movements there were some amphibians but it is after the movements that we have the great age of reptiles and the seas became populated by the well-known coiled ammonites and innumerable brachiopods. The curious graptolites which scarcely survived the Caledonian movements had completely gone. The doom of the heavily armoured and small-brained Jurassic and early Cretaceous reptiles was sealed well before the earliest inklings of the Alpine movements. The victory of the mammals, with man himself to follow later, was assured long before the time the Alpine storm really broke.

We are now in a position to apply the general principles already enunciated and to see how far they explain the building up of the

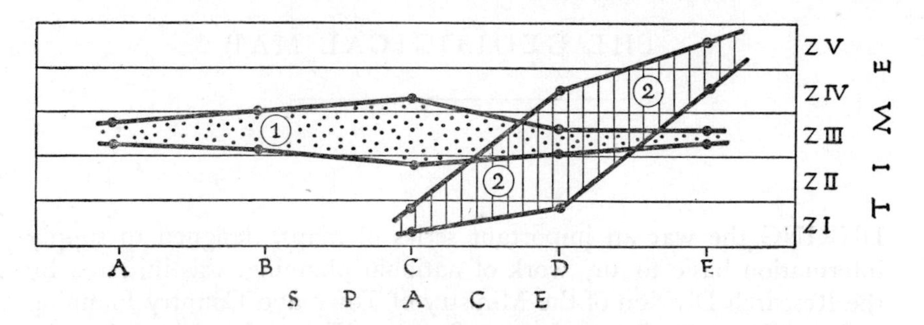

FIG. 10.—Diagram illustrating the distribution in time and space of a typical fossil. ZI, ZII, ZIII, ZIV, ZV are zones

The species 1 is found at all five localities A, B, C, D and E and is restricted in its vertical distribution to Zone III. Only in one locality, C, is it found slightly below and slightly above the limits of the zone. It is therefore a good "zonal index." Species 2, on the other hand, is only found in three of the five localities; it has a wide range in time and is found at a much lower horizon in locality C than in locality E. It is therefore of no use as a zonal index but such a distribution is characteristic of a "facies" fossil—a species seeking some special habitat conditions such as shallow water near a shoreline.

present-day structure of the British Isles and to see what evidence the rocks of this country contain of the evolution of the great world groups of animals and plants and, in the later stages, of our own particular native fauna and flora.

CHAPTER 4

THE GEOLOGICAL MAP

DURING the war an important series of maps, designed to supply information basic to any work of national planning, was initiated by the Research Division of the Ministry of Town and Country Planning and published by the Ordnance Survey. The main series is on the scale of 1 : 625,000 or approximately 10 miles to the inch and the series to be issued on that scale, covering the island of Britain in two large sheets, includes many of the maps which were proposed in the scheme for a National Atlas drawn up by the National Atlas Committee of the British Association for the Advancement of Science. Amongst the maps actually prepared and available, those of the Relief of the Land, Land Utilisation, Types of Vegetation, Types of Farming and Land Classification are very relative to the study of the natural history of this country, but fundamental to any study are the geological maps. Following the practice of the Geological Survey in some of their detailed maps, there are to be two maps. That published shows the " solid " geology as it would appear if superficial deposits such as boulder clay, glacial sands and gravels and clay-with-flints (see page 140) were removed. The other will show the " drift " geology with all those surface deposits indicated. It has already been pointed out that many geologists are interested primarily in the older rocks and in the structure of the earth's crust and the superficial deposits are to them simply a nuisance; from the economic point of view in the investigation of mineral resources the same is true and so the drift deposits and their mapping have been relatively neglected—notwithstanding their supreme importance in agriculture.

The maps published for the Ministry of Town and Country Planning[1] do not cover Ireland. In that island the surface geology is especially important and reference should be made to the " Map showing the Surface Geology of Ireland " by Sir Archibald Geikie (Bartholomew) Scale 10 miles to one inch. The Geological Survey came into existence in 1835 as an offshoot of the Ordance Survey and its immediate

[1] Now merged with the Ministry of Housing and Local Government.

task was the mapping of the geology of the country on the scale of one inch to one mile. This work was vigorously prosecuted, especially under the energetic guidance of Sir Roderick Impey Murchison who became Director-General in 1855, and the results were published in the form of hand-coloured one-inch maps, the base maps used being the original one-inch sheets of the Ordnance Survey. These original one-inch maps were " solid " maps though indications of the presence of drift is conveyed by words printed across the map in appropriate places. Its preliminary work finished, the Geological Survey then set to work to carry out a detailed revision or second survey. The field work was done on the six-inch or even, in cases, a larger scale. Attention was concentrated on the coalfields and on areas around populous centres, together with selected tracts in different parts of the country. For purposes of publication the " small-sheet series " of one-inch Ordnance Survey maps was chosen and these have been kept up to date for use as Geological Survey base maps long after they have been superseded by other editions for ordinary uses. They cover an area of 18 miles from east to west and 12 miles from north to south, and have the advantage of showing contours in black as part of the base map. Some of the small sheets series are hand coloured but the technical advances in colour printing have meant that all the later maps have been printed in colours. Each sheet is accompanied normally by a detailed explanatory memoir. In those areas where drift deposits are widespread or important it is usual to publish two editions of the map, the Solid and the Drift Editions, which are in fact two distinct maps. After more than half a century of work only about a third of the country has been re-surveyed and the maps published, so that for the rest reliance still has to be placed on the original survey and hand-coloured maps of a century ago. There is thus an obvious difficulty in issuing generalised maps of the whole country in that the detail available is so varied from one part to another, and it explains why, in a country where superficial deposits are of such tremendous importance in the study of soils, vegetation and agriculture, there is no generalised map to show their distribution. In addition to the map of solid geology on the 10-mile scale just mentioned there is a useful map, also of the solid geology, of the whole of the British Isles on the scale of 25 miles to one inch, published by the Ordnance Survey as one sheet at a very modest price.

This map is of the greatest value in giving a general picture of the distribution in Britain of the rocks of each of the systems. In the third edition, which is dated 1939, very considerable revisions and additions were made as a result of incorporating recent work.

Speaking very generally, in geological terms the oldest part of Britain is the north-west and on the whole the rocks become steadily younger in geological age as one goes towards the south and east so that the major stretches of the Tertiary rocks are to be found in the London and Hampshire Basins.

The great mass of the Highlands of Scotland is made up of a complex of ancient metamorphic rocks with numerous large intrusions of granite. As described more fully in Chapter 23, one must picture the whole as the worn-down stumps of the great Caledonian fold mountains and there is little to-day in the relatively tame, rolling relief of much of the Scottish moorland to suggest the wild contortions exhibited by the underlying rocks—structures the interpretation of which has long baffled and continues to baffle the most expert of geologists. The oldest rocks of all are probably the Lewisian gneisses in the Outer Hebrides but the relative ages of the different pre-Cambrian or Archaean rocks is still uncertain. Along the coastlands of the North-West Highlands is a considerable stretch of Torridonian Sandstones—still pre-Cambrian but unmetamorphosed sediments, obviously much younger than the main bulk of the Highland rocks (see Plate 1B. There is a narrow belt of Cambrian also in the north-west of the Highlands but the main masses of Old Red Sandstone—reminders of the great lake basins created by the Caledonian upheaval—lie on the east. Tiny patches of Jurassic rocks both in the west and along the east coast in the far north are reminders that Jurassic seas must have stretched far north but have left only small traces, and there is little evidence of the detailed geological history of Scotland over vast periods of time. Though the Alpine earth movements failed to fold the rigid old mass of the Highlands there is abundant evidence of the way in which great cracks were formed through which poured masses of molten lava. These make the great red splashes on the map in Skye and Mull and many of the smaller Hebridean islands.

A glance at the distribution of the deep purple colour used to indicate the Cambrian and the mauves used for the Ordovician and Silurian serves to demonstrate that it is the Older Palaeozoic rocks which make up the Caledonian mountain ranges of the Southern Uplands of Scotland, the English Lake District and the Isle of Man, as well as the whole of north and central Wales. That the Old Red

Plate I

LANGSLEDDALE, LAKE DISTRICT

AEROFILMS

A scene typical of Highland Britain, where valleys and lowlands which can be settled and farmed are separated by wide stretches of mountain and moorland

Plate II

THE CHALK COAST OF DORSET, WITH OLD HARRY ROCK AERO-PICTORIAL
Lowland Britain is far from being a uniform plain but few parts are too elevated or too poor to be farmed

Sandstone occupied basins is not only clear from the Highlands of Scotland: there are broad belts in the great tectonic depression of the Central Lowlands of Scotland between the Highlands and the Southern Uplands. In the south the main stretch of the Old Red Sandstone is in the Welsh Borderland whilst marine Devonian rocks of the same age cover Exmoor and much of South Devon and Cornwall.

Three colours are used for the rocks of the Carboniferous system—blue for the Carboniferous Limestone and the Scottish rocks (with sandstones, shales and coals) of the same age; sage green for the Millstone Grit and barren Culm Measures of Devon and Cornwall, and slate-grey for the Coal Measures. In passing it must be noted that the outcrops of Coal Measures shown on the geological map are not co-extensive with the coalfields because much of the most valuable parts of the coalfields are hidden beneath younger rocks. The Carboniferous rocks, with many areas of lava and other reminders of volcanic activity, fill in the remainder of the Central Lowland of Scotland whilst in England one is struck at once by the great north-south stretch of Carboniferous rocks which makes up the Pennines—the so-called backbone of England. This north-south alignment is a new one: it is a reminder that the rocks were deposited long after the Caledonian folding and that after their deposition were subject to the Armorican movements. These created north-south folds such as the Pennines and Malverns, folded as it were against the older blocks to the north and west, as well as the more common east-west folds so well seen in the alignments of the beds in South Wales and in the South-western Peninsula. Where the north-south and the east-west folds crossed, the creation of basins and dome-shaped uplifts is obvious and can be clearly seen from the map. Thus the South Wales Coalfield and the Forest of Dean are examples of basins and the Mendips are an example of the uplifts.

The way in which the bright blue streak of the Magnesian Limestone (Permian) cuts across different beds of the Carboniferous shows that the latter had already been folded and denuded before the deposition of the Permian.

The remainder of the map relates to Lowland Britain. The Midlands of England show clearly the stretch of the red Triassic deposits and the islands of older rocks which appear from beneath this cover. Then follow the successive belts of the Scarplands (see Chapter 18), sweeping across England from north-east to south-west—brown for

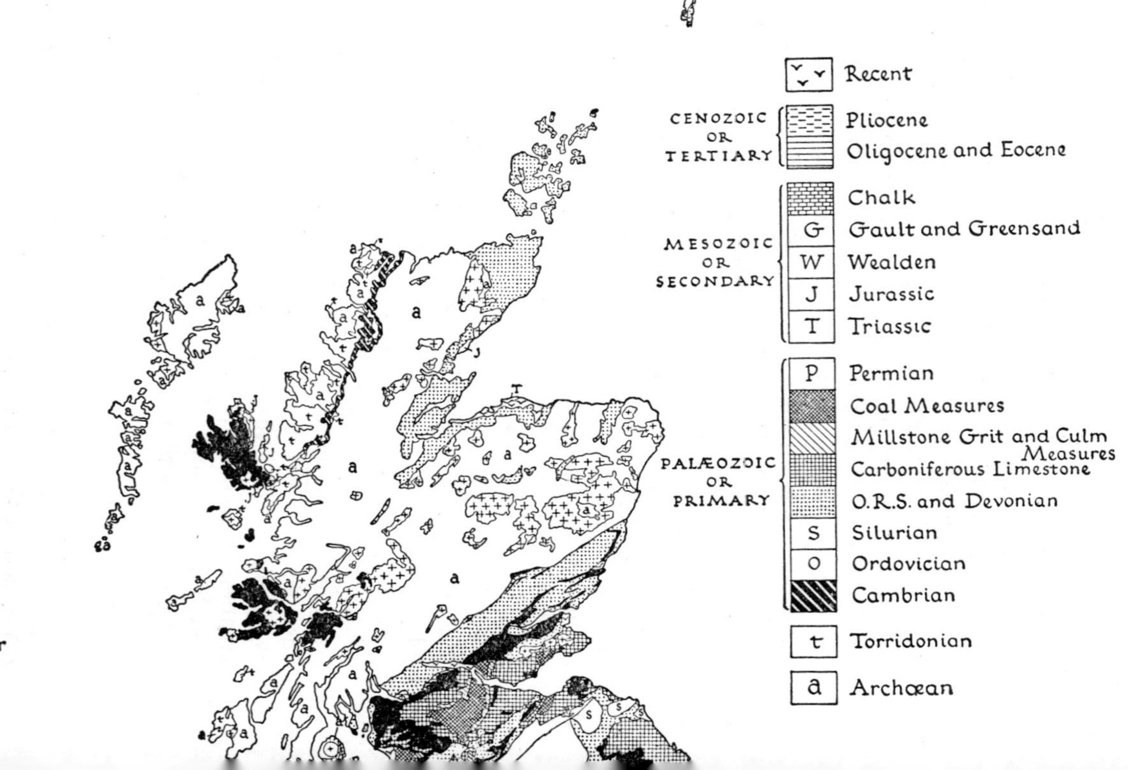

Recent
CENOZOIC OR TERTIARY
Pliocene
Oligocene and Eocene
MESOZOIC OR SECONDARY
Chalk
Gault and Greensand
Wealden
Jurassic
Triassic
PALÆOZOIC OR PRIMARY
Permian
Coal Measures
Millstone Grit and Culm Measures
Carboniferous Limestone
O.R.S. and Devonian
Silurian
Ordovician
Cambrian
Torridonian
Archæan
Volcanic rocks of various ages

Granites and other intrusive rocks of various ages

FIG. 11.—A Simplified Geological Map of the British Isles

the Liassic clays, yellow for the Oolitic sequence, dark green for the lower Cretaceous rocks, and light blue-green for the Chalk follow one another in sequence of ever decreasing age as one goes towards the south-east. It is in the south-east that the orderly sequence is interrupted and this is a reminder that the south and south-east of Britain lay on the fringes of the great Alpine storm and that some of the rocks there were folded by the Alpine movements. The uplift of the Weald, roughly east and west in its main axis, separates the two main Tertiary basins of London and Hampshire ; the sharpest folds of Alpine date are those in the extreme south—across the Isle of Wight and the Isle of Purbeck. The location of the main stretch of Pliocene rocks in the coastal parts of Norfolk and Suffolk is suggestive, and rightly so, that by that period the geography of Britain had acquired something of its present form—only certain parts, on the whole near the present coasts, came under Pliocene seas.

There the geological story shown by the general map we have been discussing ceases. It tells us nothing of the stupendous events of the Great Ice Age. For that we must turn to the detailed drift maps and from them try to piece together what is one of the most fascinating and important, and yet, despite its recent occurrence in terms of geological time, one of the most difficult episodes to reconstruct in the geological story of the building of Britain. Before we deal with this, however, it is essential to consider how many of the major surface features of Britain have evolved.

THE WHALE ROCK, BUDE, NORTH CORNWALL L. D. STAMP

Beds of sandstone have been folded into a small arch or anticline or, more strictly, a pericline, because it is "pitching" at both ends (towards the reader and, at the far end, towards the sea)

L. D. STAMP

FOLDED ROCKS EXPOSED ON THE FORESHORE, NORTH OF BUDE

On the left is a syncline, on the right an anticline, both pitching seawards

Plate IV

STRATA CLIFF, MILLOOK, NORTH CORNWALL L. D. STAMP
Highly folded beds of sandstone

BRITISH ASSOCIATION
AN UNCONFORMITY AT HORTON-IN-RIBBLESDALE, YORKSHIRE
The upper beds, belonging to the Carboniferous Limestone, are seen resting almost horizontally on the upturned edges of the underlying Silurian shales

LAND FORMS AND SCENERY

The Work of Rivers

ALTHOUGH it is clear enough that the form of the surface—the relief of the land—is in the main determined by the underlying geological structure, the relationship between land forms and structures is by no means a simple one. It is only within the past few decades that the specialist study of land forms, the science of geomorphology, by developing its own technique, has demonstrated that it is possible to reconstruct a long and often complex history by detailed investigation of the form of the ground and that the details of land relief may bear surprisingly little relationship to the structural geology. For the most part geologists have paid little attention to this natural development of their studies. Apart from a few outstanding works by geologists such as J. E. Marr's *Scientific Study of Scenery* first published in 1900 and more general works such as Lord Avebury's *Scenery of England and Wales*, the foundation of detailed work was laid by such physical geographers as the American W. M. Davis, whose famous studies of the evolution of rivers was nevertheless carried out in our own Wealden country, and the Frenchman Emmanuel de Martonne whose *Traité de Géographie Physique* contains many British examples. Much recent work has emanated from America and other detailed work from Germany. In this country some of our leading geographers, headed by Professor J. A. Steers and W. V. Lewis, have concerned themselves especially with coastal phenomena, others such as Professor D. L. Linton and Professor A. A. Miller with river evolution whilst a leader amongst those devoted to general geomorphological studies was Professor S. W. Wooldridge, whose *Physical Basis of Geography: an Outline of Geomorphology*, was first published in 1937.

Briefly, it may be said that land-forms depend first on the nature of the rocks and their disposition (that is, in other words, on lithology and structure), secondly on the climatic conditions, with resulting soil mantle and vegetation cover, under which the sculpturing of the land surface has been and is taking place, and thirdly on the phase or stage within the erosion cycle (see above, pp. 21-22).

However erroneous, it is common to find references to " hard " rocks and "soft" rocks which are regarded as respectively resistant to and less resistant to weathering. Since most of the older rocks are "hard" in this sense the common distinction is drawn between the old hard rocks and the young soft rocks characteristic respectively of Highland and Lowland Britain. Although in any given area it is broadly true that the positive features of the relief, the mountains, hills and plateaus, are coincident with the outcrop of resistant rocks and the negative features, the valleys and plains, to that of "weak" rocks, resistance to weathering is not a matter of actual hardness. Chalk could not be described as a hard rock, yet it gives rise to the main hill ridges of south-eastern England. Under certain circumstances even a bed of gravel is sufficiently "hard" to form a capping and preserve a hill from denudation as in the case of Shooter's Hill to the south-east of London. Both with chalk and gravel this is largely due to the fact that rain water soaks *into* the rock so readily that it does not have time to collect in rivulets *on* the surface and wash away the surface soil. When reached in deep excavations such as wells even clay is quite hard but when at the surface it has absorbed a certain amount of water it is impervious to more. When rain falls on the surface it is then easily eroded—as muddy streams bear witness—and so outcrops of clay are marked by valleys and lowlands.

In the British Isles we are concerned with the land-forms which develop in a moist, temperate climate. We are not, for example, directly concerned with land-forms which develop in hot deserts or in the rainy tropics except in so far as such conditions once prevailed in distant geological epochs and have bequeathed to us fragments of "fossil" landscapes in the sun-shattered rocks which peep from beneath a cover of later strata in the Wrekin or the ridges of Charnwood Forest to remind us of the deserts of Triassic days. We are, however, concerned with land forms which develop under conditions of extreme cold under great ice-sheets or valley glaciers or on the margins of ice-covered seas, for much of the surface of this country was profoundly modified during the Great Ice Age. This is geologically so recent that not a few of our lakes and swamps are the last remains of those left behind by the retreating ice.

Over large parts of this country the relief seems to be completely unrelated to the underlying structure. Plains are developed quite independently of either the hardness or dip of the underlying rocks : rivers seem to go out of their way (as does the Bristol Avon) to pass through the highest hill ranges they can find instead of following an easy passage on low ground and it is here that we realise the importance of the erosion cycle. It is in the interpretation of such apparent anomalies that the geomorphologist has made his major contribution. In the following pages we shall examine in detail a number of examples from Britain.

The Work of Rivers

The principal agent in the sculpturing of the land surface in a rainy temperate climate such as that of Britain is undoubtedly running water. No sooner does rain fall than some of it collects to form tiny temporary rivulets which soon join small permanent brooklets and rills. These, reinforced by springs which represent the reappearance at the surface of that portion of the rainwater which had soaked into the ground, unite in due course to form the river system of the country. Except in certain limestone districts where much of the drainage is underground the whole country is covered with a complex surface drainage pattern of rivers and streams.

It is a common generalisation in most books on physical geography that the course of a river may be divided into three parts. The upper or mountain course is that in which swiftly flowing water, especially after rain, is able to move stones of considerable size, to roll them along, to rub them one against the other and so to reduce angular fragments such as those broken off the mountains by frost action into rounded pebbles. The work of such a mountain torrent is well seen in Plate 2. At the same time the river deepens and widens its own valley so that its valley has a typical V-section with unstable banks. The middle course of the river is that over the foothill belt where it has lost some of its velocity but is still moving rapidly enough to carry sand, silt and mud in suspension and to roll pebbles along its bed. Its main work there is transportation ; its valley has a broad open section and has stable sides so that the erosive power of the river is strictly limited. The lower course is that in which the river meanders lazily across a plain ; though sweeping much mud out to sea or into the lake into which it empties it has lost much of its velocity and so much of its power of transportation. It lays down part of its load as shingle beaches or sandbanks but especially builds up large flat plains of deposition by spreading alluvium over a wide flood plain or a delta. Thus its work is largely deposition.

Quite obviously not all rivers conform to such a generalised pattern. In mountainous regions they may tumble direct from the mountains to the sea (as shown on Plate XXVIII)—they are young rivers associated with an early stage in the cycle of erosion. Others, including such large rivers of this country as the Thames, have no mountain course—they are relatively mature and associated with a late stage in the cycle of erosion.

It is clear that there is a very close relationship between the

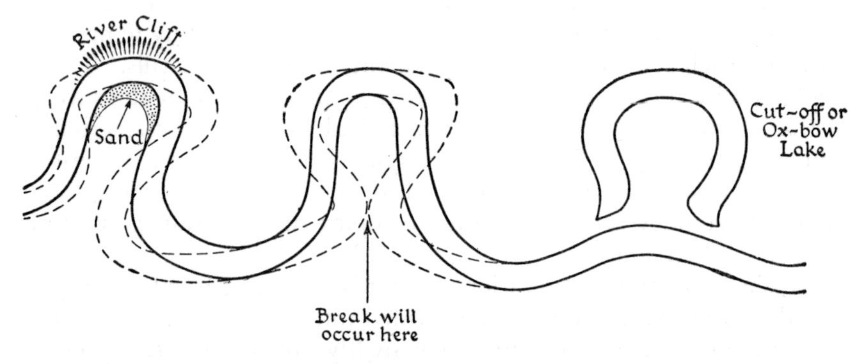

FIG. 12.—Diagram of a Meandering River

This diagram shows how the water of even a slowly-moving river in its lower course swings from side to side resulting in erosion on the concave side and deposition of sandbanks on the convex side.

character of a river and the phase of the erosion cycle. It is the inexorable law of nature that as soon as mountain building movements have erected a land mass and endowed it with mountains, hills and valleys, the forces of sub-aerial denudation combine to reduce that land to . . . the question is to what? What is the final form of the land if the forces of denudation are allowed to continue unchecked? The answer is not a flat plain but a peneplain or peneplane which, whichever way it is spelt, means almost a flat surface. The whole process is referred to as "sub-aerial peneplanation." Although sub-aerial peneplanation is in progress all over the world and although large areas may thus have been so reduced that they have reached "base level", below which removal of material will not take place, it is rarely if ever in nature that the process is allowed to continue to its logical conclusion. Differential uplift of the land, up or down movements relative to sea level (eustatic movements), or even a slight folding movement will upset the equilibrium which has been reached and will cause "rejuvenation" of the river systems. Before considering these complications it will be well to note the various ways in which river systems may develop and thereby to explain many of the features which are associated with British rivers. In passing we may recall that peneplanation of lands in the British Isles has probably been almost reached in past geological epochs. The formation of shallow water

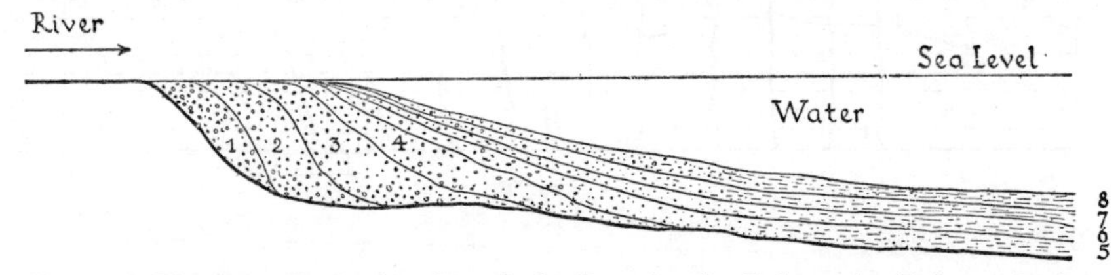

FIG. 13.—Diagrammatic Section through the Deposits of a Delta and a Lake or the Sea

When the river enters a lake or the sea, the velocity of the water is immediately checked and any coarse material is at once dropped, only the finer mud being carried on. In this way a delta is built up and as the flat land of the delta itself is formed, the velocity of the river is checked before reaching the sea and the finer deposits are spread as a surface layer of alluvium over the flood plain.

limestones which requires clear water is taken as evidence that the surrounding lands had been almost reduced to base level and consequently yielded very little sediment.

We notice that there are thus two types of plains—plains of deposition and plains of denudation and that the former tend to be flatter and are more truly plains.

Let us take the simple case of the floor of the sea which is raised up (not folded) by earth movements so that it becomes land. It will be a flattish surface with a gentle slope seawards and rain falling will collect together into streams, roughly parallel, finding the shortest route seawards. These streams are consequent on the slope and hence are known as *consequent* streams. Tributary streams, arranged somewhat irregularly, will drain into these main ones and the pattern of drainage developed is that known as dendritic (Greek, *dendron*, a tree).

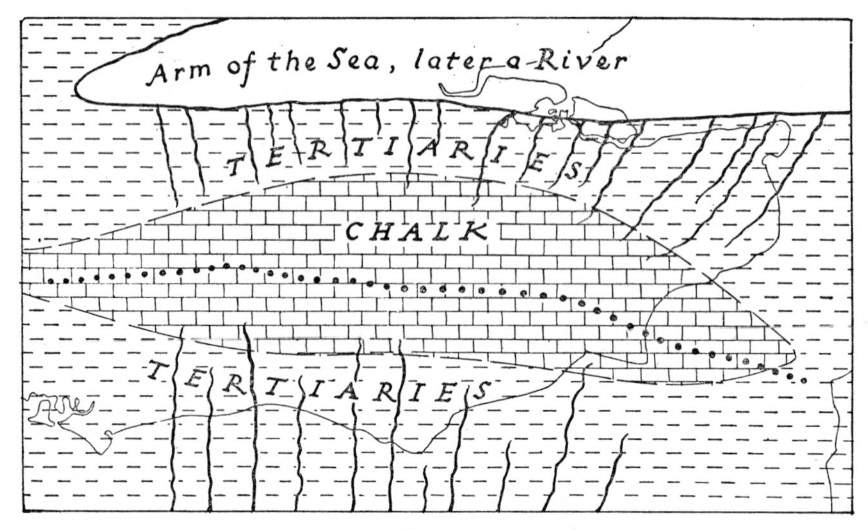
Arm of the Sea, later a River
TERTIARIES
CHALK
TERTIARIES

FIG. 14

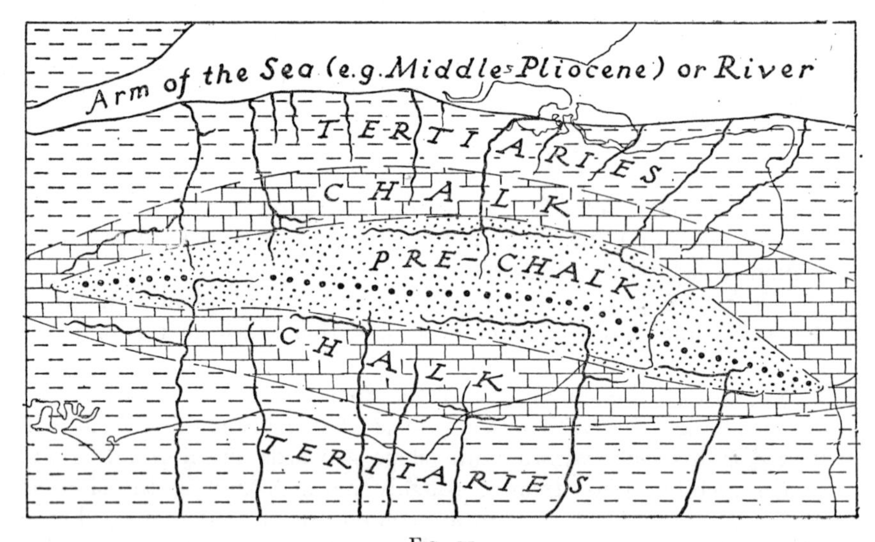
Arm of the Sea (e.g. Middle Pliocene) or River
TERTIARIES
CHALK
PRE-CHALK
CHALK
TERTIARIES

FIG. 15

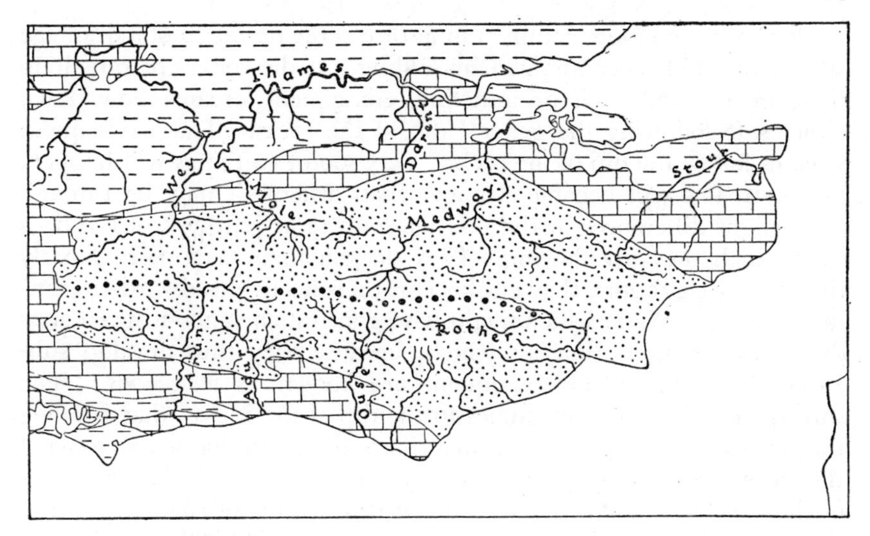

FIG. 16

Figs. 14 to 16. These illustrate three stages in the development of the drainage of the Weald. In each the line of dots represents the main axis of the Wealden uplift. When the dome was first uplifted (Fig. 14) chalk covered the whole and water drained naturally and consequently *down* the northern and southern slopes forming *consequent* streams. In the next stage (Fig. 15) the Chalk has been removed by denudation over the central area and some streams have become stronger than others and *subsequent* streams, running in strike valleys, have developed. Fig. 16 shows the developments at the present day. The three divisions shown are the Tertiary, the Chalk and the pre-Chalk beds. It is clear that such a river as the Darent has had its headwaters captured by the Medway.

Perhaps even more common in nature is the initiation of a drainage system by uplift accompanied by folding. If the rocks are raised up to form a broad arch or anticline, consequent streams flow down either side following the general dip of the rocks. Thus consequent streams follow the dip of the rocks. Very soon two things will happen. Some streams will become stronger than others—it may be through some slight differences in the relative softness or hardness of the beds over which they are flowing. They deepen their beds more rapidly than their neighbours, they cut back at their heads (headward erosion) more rapidly. Water which might have gone into neighbouring streams drains into them by laterals which, because they thus develop subsequently to the main consequent are known as *subsequents.* The subsequents flow at right angles to the dip of the rocks—that is along the strike—and so are flowing in strike valleys. In due course some of the more vigorous subsequents capture the waters and drain the valleys of the weaker consequents and so, by this process of river capture, a complex system develops. It may even happen that the flow of water in part of a former consequent valley is reversed so that it becomes an *obsequent* stream feeding the conquering subsequent.

Such a river system as that just described may cut down on to older rocks which lie underneath the sheets of strata which gave it its birth. Indeed all traces of those later rocks may be completely removed. On to the older rocks there is implanted a river system which seems to have simply no relationship to the structure. This is a common feature in many parts of the British Isles and gives us the reason for the passage of the Bristol Avon through Clifton Gorge when there apparently were so many other easier courses. Such a system is called a superimposed drainage. But in its further development the members of the system will find out the weaker rocks, the lines of faulting and crushing, and it is along such lines that the major excavation will take place.

What happens when earthquakes and folding movements take place in an area with a well developed river system? There may be reversals of drainage and many examples of river capture can only be explained by postulating differential earth movements. But if folding movements take place slowly existing rivers in their downcutting may keep pace with the growth of folds and one gets thus examples of antecedent drainage which may be defined as drainage developed in its early stages *before* the present surface features. For the supreme

THE LANDSLIP, ST. CATHERINE'S POINT, ISLE OF WIGHT AERO-PICTORIAL

The cliff behind consists of Lower Chalk and Upper Greensand resting on Gault clay (hidden), the whole dipping gently seawards so that the upper beds have slipped over the surface of the clay. The whole of the foreground consists of slipped rocks

Plate VI

HONEYCOMB WEATHERING OF SANDSTONE L. D. STAMP

Originally a bed of soft sand, the grains have been cemented together by iron oxides deposited from solution both along certain bedding planes and along vertical cracks. The unconsolidated sand may later be washed out by wind, rain or (as in this case) by the sea

examples of this one must look to the mighty rivers of the Himalayas which cut right through the greatest chain of mountains on the face of the earth.

The sequence of development of consequents, subsequents and obsequents was applied by W. M. Davis to the river systems of the Weald which is thus classic ground. One may picture the Wealden dome, in structure resembling an overturned boat, rising by slow stages from the latter part of the Cretaceous period downwards. At first the uplift formed a low dome scarcely above sea level but sufficiently near the sea surface for wave action to get to work wearing away the chalk and rolling the angular flint nodules into pebbles. By Middle Eocene times the ridge was sufficiently high to be partly covered by shingle beds (Blackheath Pebble Beds) and the crest probably formed an island. As soon as an island appeared above the surface consequent streams flowing from the east-west crest to the north and to the south developed. By a combination of marine erosion and then of sub-aerial denudation the chalk was entirely eroded from the central area and revealed below the varied succession of beds which make up the lower Cretaceous. Some of the beds are weak and easily eroded, others are relatively resistant. Subsequent streams found out the weaker rocks and eroded valleys at right angles to the consequent streams, some of which cut down through the chalk and to-day are seen flowing in steep-sided valleys through the chalk rim of the North Downs and the South Downs. The weaker consequents were beheaded by the capture of their head streams and many failed to cut down through the chalk. The Weald thus illustrates extremely well the association of subsequent streams with valleys in the weaker rocks which are parallel to the strike of the rocks (strike valleys) whereas the consequents have valleys parallel to the dip of the rocks (dip valleys). The later history of the Weald has been complicated by the submersion of much of the area under the Pliocene sea, then its subjection to tundra conditions during the Great Ice Age (see page 155) and by the complications caused by the breaching of the eastern end of the fold when Britain became separated from the continent, but the main pattern of the drainage has remained as it was developed by the gradual uprise of the Weald. The phenomena of subsequent streams occupying well-defined strike valleys is repeated all over the lowland of Britain.

The form of a river valley is able to yield much information both with regard to the age of the valley itself and the history of the river system.

Mountain torrents stand rather by themselves: they cut deep notches in the mountain sides (an example is given in Plate 2), usually finding some line of weakness, as for example along a fault where the rocks have been crushed, and the material which is dislodged is swept to lower levels both by the power of the water and by the force of gravity. If dislodged blocks fall by gravity alone they form screes with an angle of rest of about 40°—the angle of the scree shown on Plate 8B is exactly 38°. If the fall is aided by running water the debris is fanned out and has a lower angle of rest—forming what is termed an alluvial fan or alluvial cone (such as the ones shown on Plate 30B) though the word "alluvial" is apt to cause confusion with the much finer material associated with deltas and with the flood plains of the lower courses of rivers.

Where initial slopes are not quite so steep the mountain stream carves out a narrow steep-sided V-shaped valley. Even at this early stage the valley is not straight: the stream swings from side to side so that "interlocking spurs" develop between the meanders and obstruct the view upstream. Nature has provided the swiftly flowing stream with a remarkable mechanism for drilling holes in its bed. A few stones are caught in a whirl of water and swing round and round to drill out the well-known "pot-holes." This is an active force in deepening the bed of the river and so of its valley. An excellent and large example is shown in Plate VIIIA. Widening of the valley comes gradually with the action of gravity—lateral slipping aided by tributary streams so that, broadly speaking, the older or more mature the valley the wider it is. In these early stages the form of the valley, especially its long section (i.e. the section drawn down the valley—the longitudinal profile for which the not very appropriate German word *talweg* is often used), is closely related to the character of the rocks over which it passes. Hard bands cause rapids or waterfalls and between these the river may assume the characteristics of maturity. In cross-section the valley sides may exhibit ledges due to the outcrops of hard bands whilst dipping strata may cause a valley with an asymmetric cross section. Even more common is the varying width of the valley—broad and open where it traverses soft rocks, narrow and even gorge-like where it passes through a belt of hard rocks or limestone. Even an old river like the Thames has these features—the beautiful narrow valley at Goring is where it passes through the chalk ridge.

Gradually, however, a river tends to reach a state of equilibrium and its longitudinal profile will form a smooth curve from source to mouth. When it reaches this stage a river is said to be graded and the land around has reached the stage of sub-aerial peneplanation. To achieve the graded curve, which will first be reached near the river's mouth, the stream must necessarily cut back into the hills from which it takes its source and this involves headward erosion. It is found that

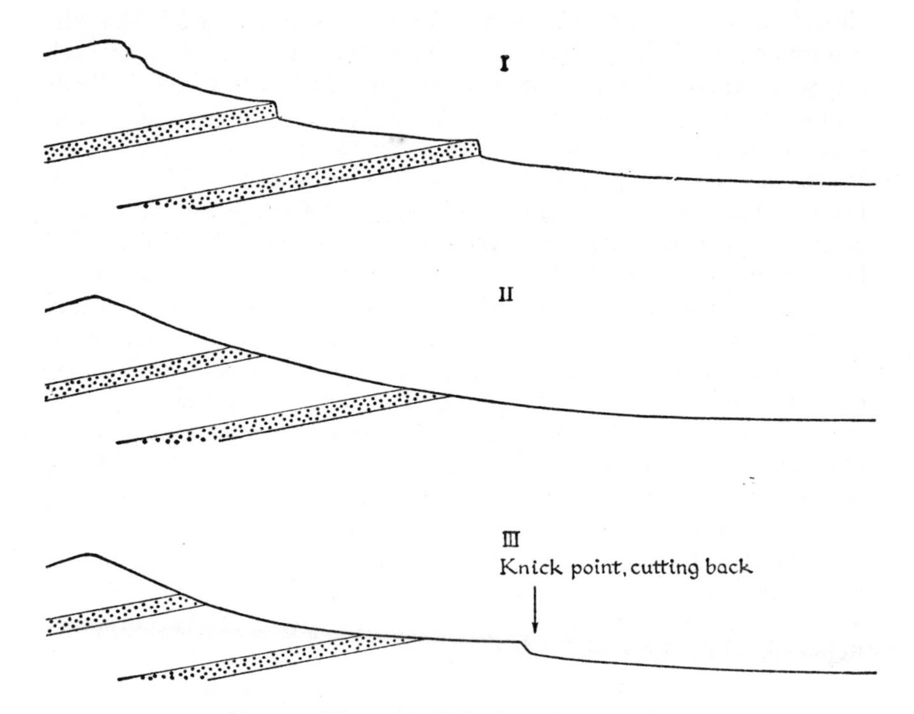

FIG. 17.—Diagrammatic Sections along a *Talweg*

The upper diagram is a longitudinal section following the course of a relatively young river from its source to its mouth. Bands of hard rock cause waterfalls and rapids between which the river tends to assume a graded curve.

Diagram II is the graded curve of a more mature river : the whole longitudinal section is evenly graded from source to mouth independently of any hard beds.

Diagram III illustrates what happens if a fully graded mature river, such as that shown in II, is subjected to rejuvenation by a general uplift of the land surface relative to sea level. A knickpoint is formed independently of the character of the rocks and gradually works back, i.e. up the course of the river.

many mature rivers rise in a sort of amphitheatre, steep-sided but not nearly so steep-sided as the cirques from which glaciers have their origin (see page 87).

Over the middle and lower courses of mature rivers, or rivers which have almost reached base-level, there are several characteristic features. The water swings from side to side and long winding meanders are the result (see Fig. 12). Once a meander has been initiated there is a natural tendency for the swing of the water to make the curves ever more acute till at last the water breaks through the neck and the cut-off portion forms a stagnant "cut-off" or "ox-bow" lake. This will be clear from the diagram; but what is not always realised is that the continuance of such a process results in a broad flat-floored valley with a deposit of gravel, sand, silt or alluvium. Such a flat floor is liable to flood when the river is in spate and so one gets a flood plain. Land liable to flood occurs along the lower courses of most British rivers. When the flooding is uncontrolled, a film of mud is spread by each flood and results in the gradual building up of alluvial flats. There is thus deposition closely associated with erosion in the middle and lower courses of a river. When the river reaches its mouth with a load of fine mud in suspension this may be swept seawards, especially if the sea into which the river discharges has a marked tidal movement. This is the case round the British Isles where nearly all our rivers enter into estuaries with a strong tidal movement. Where tides and currents are less strong the sediment is dropped near the mouth of the river and a delta of alluvium is gradually built up, passing seawards almost imperceptibly into very shallow muddy water. Since deltas are not typically formed round Britain it is unnecessary to enter into the details of their formation though there are many good examples where rivers enter lakes such as that shown in the foreground in Plate 31B. It is important to note the leading role played by vegetation in fixing the mud and then acting as a trap to catch more mud. In this way, though not directly associated with river mouths, there is accretion of land in such areas as around the Wash and in Morecambe Bay and advantage is taken of the natural processes in reclaiming land by building dykes or retaining walls to hold sediment. The stages in silting up are well shown in Plate XXV. Inland, artificially controlled flooding has long been practised, using the waters of such rivers as the Trent and Yorkshire Ouse to spread silt over the land after the manner of the Nile in Egypt and so both to build up the level and to spread a

Plate VII

THE ROCKY BED OF THE RIVER USK, NEAR BRECON
Taken at a low-water period, showing the scouring of the rocks by "potholing"

L. D. STAMP

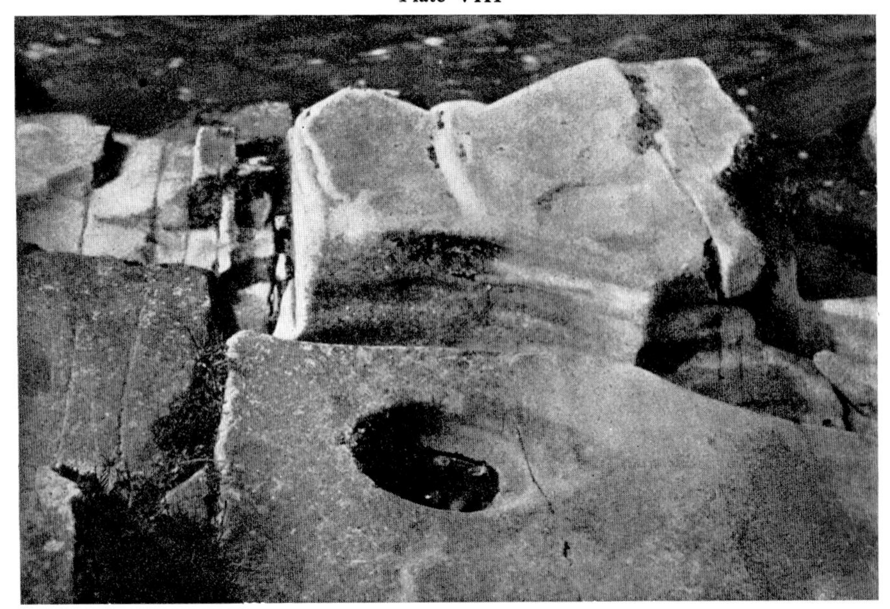

DETAILS OF A POTHOLE, RIVER USK, NEAR BRECON L. D. STAMP
The stones which are whirled round when the river is in flood and so bore out the pothole can be clearly seen

HEAD OF WASTWATER, LAKE DISTRICT L. D. STAMP
Much destruction is caused in times of flood, when the mountain torrents sweep down masses of coarse, roughly-rolled pebbles even over former meadows, as in the foreground of this picture

fertile layer rich in mineral salts and organic matter and of excellent mechanical texture. This controlled flooding is known as warping and the mud deposited as warp.

The well-graded meandering river with its broad valley floored with alluvium is a familiar feature in the British landscape. But even in geologically recent times, certainly since the Ice Age, there have been several changes in the relative level of land and sea, slight it may be but significant. What happens to such a mature river system when the land is lowered or raised relative to sea-level ? First, if the land sinks, the lower valley is invaded by an arm of the sea and one gets the familiar feature of a drowned valley or *ria.* The best example of a coastline of drowned valleys or ria coast is the south-west of Ireland. Soundings show that the floor of the ria, the old river talweg, slopes steadily seawards and there is no " lip " as there is in the case of a glaciated valley with a rocky or morainic bar at the entrance (as in many of the Scottish fiords, see page 234). Drowned valleys give rise to the picturesque winding creeks of south Devon and Cornwall—the estuary of the Fal and Tamar for example (see Plate 26). It is clear that the branching tidal creeks shown in Plate 26 could not have been excavated by the action of the sea which now occupies them.

If, on the other hand, the level of the land is raised relative to the sea, the river undergoes rejuvenation ; it is given new erosive powers and immediately begins lowering its bed. But such a rejuvenated river exhibits certain special features. It was, before the new uplift, a meandering mature river and the effect of the uplift is for it to follow its same meanders but to cut them deeply and so one gets the interesting and picturesque feature of *incised meanders* with a river winding in a gorge, it may be of considerable depth. If in such a case a meander is cut off one gets between the abandoned course and the new course a " meander core." Incised meanders tend to develop where the rocks are relatively hard. Where a broad valley is excavated in relatively soft rocks the rejuvenated river develops for itself a new alluvial covered flood plain at a lower level than the old one and so fragments of the old one are left as gravel-covered or alluvium-covered terraces. Successive uplifts produce successive terraces at several levels. Those of the Findhorn in Scotland are well shown in Plate XXXI. The terraces of the Thames are not only well known but have been and are very important economically—for the dry sites they offer for settlement, for the water supplies once afforded by the gravels, for the excellent well-drained soils to which they give rise, for the brickearth they formerly supplied for brick making, and latterly for the supplies of gravel which, alas, is being excavated regardless of the future use of the devastated land. In the case of the Thames near London it is possible to distinguish one gravel-covered terrace at about 100 to 120 feet above present sea-level, though naturally varying in height with distance from the sea. This is the Boyn Hill Terrace and is very clearly marked in several areas. There is another terrace, of wide extent, at about 50 feet above sea level known as the Taplow or Middle Terrace. A third one is the Low or Flood Plain Terrace at some 10 or 15 feet above sea level. Then there followed a time when the Thames was lower than at present—or rather when the sea-level was lower and the river excavated what is now a buried channel so that to this extent the estuary of the Thames is a drowned valley. Actually the history of the Thames is much more complex than this, and such a complex history is typical of British rivers. Each change has some corresponding effect on tributaries. In the lower courses of a well graded river the effects of hard bands which may cross the valley have been eliminated and an interesting feature is found when the course of a rejuvenated river is followed upstream. There is found to be a point where there is a break in the longitudinal profile of the river. This is where it is still cutting back as a result of the change in level. Such a break of slope is known as a " knick point " and its development is to a large extent independent of any differences in the rocks of the river bed.

It must be remembered that the British Isles had a well developed river-system before the oncoming of the ice sheets of the Great Ice Age and that the effect of glaciation was to modify rather than to change completely the existing valleys and land forms.

A number of the plates in this book illustrate a few of the extra-ordinary complex character of British rivers. A drowned estuary such as that shown in Plate 26 may become silted up and a marshy plain may result—well seen in the estuary of the Glaslyn in Plate 20. Rejuvenation may result in gorges even in the middle courses of rivers—as shown in Plate 2. A mature landscape with well-rounded hills affected by rejuvenation is often more apparent from the air than on the ground and an example from the Southern Uplands is well shown in Plate XXII.

CHAPTER 6

THE WORK OF THE SEA

THE extraordinarily varied character of the sea coasts of Britain and the variety of habitats which they afford to both plants and animals, with the consequent enrichment of our fauna and flora, give a special interest and importance to the story of the work of the sea in the building of the British Isles.

It is now generally agreed that ocean currents play but a very small part in the erosive and transporting work of the sea and that the effects of tidal movements are limited to a few special cases—notably tidal scour in confined estuaries and straits. The work of the sea is primarily through wave action—to some extent through the hydraulic forces engendered by the movement of great masses of water, but far more through the arming of the waves with quantities of rocks, stones, gravel and sand.

The waves of the sea are primarily wind-waves ; they are caused by the disturbance of the surface by wind but, once formed, waves may travel far beyond the area where they were generated—hence " swell " or " ground-swell " unaccompanied by wind. It is, of course, well known that there is no forward movement of the water in wave action, except where the waves are breaking on the coast. The vertical range of motion, in other words the height which waves may reach, is commonly much exaggerated. Waves which are as much as 50 feet from trough to crest are decidedly large, probably quite exceptional even in the open ocean. At a depth of 100 feet the water is little disturbed, at a depth of 500 feet it is doubtful whether there is enough movement to disturb even the finest mud. There is thus a fundamental difference between sub-aerial denudation, which takes place at all heights from sea-level to the tops of the highest mountains, and marine denudation which acts on a very restricted vertical plane above or below the surface level of the water. The maximum effect is where sea meets land—between the tide marks and just above or below.

Consider what happens at the base of cliffs. Angular blocks of rock and stones fallen from the face of the cliff are picked up by the waves and hurled against the *base* of the cliffs which they thus tend to under-

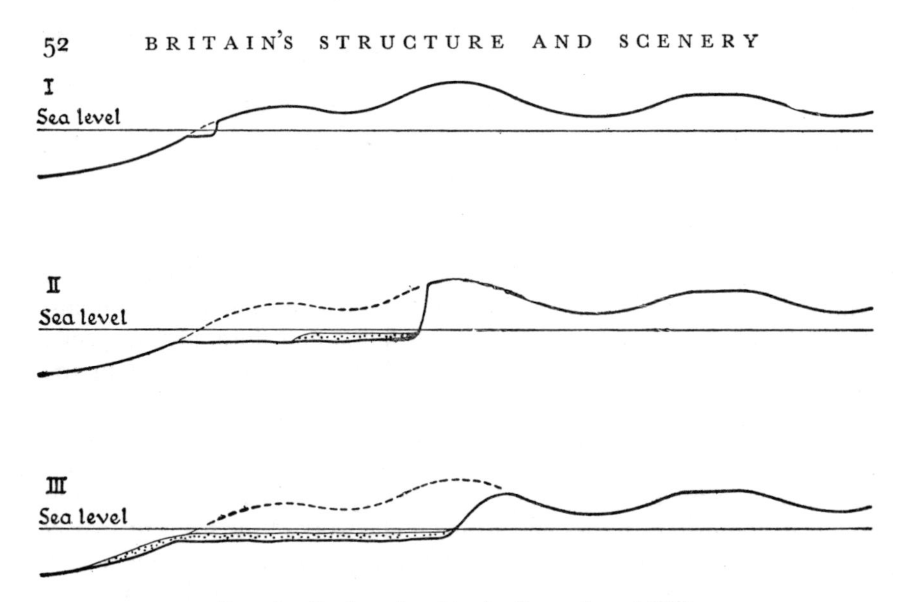

FIG. 18.—Sections showing the Formation of Cliffs

These sections illustrate the plane of marine erosion (see Plates 3 and XX A) and the way in which the cliffs are cut back and a submarine peneplane formed.

cut, much in the manner of coal-cutting machinery at the base of a coal seam. Blocks from above then split off along joints and fall by the force of gravity; where there is a dip of the rocks seawards great masses may slide down the bedding planes. The latter effect is well seen where rock overlies clay the surface of which becomes slippery and acts as a greased plane—hence the constant slips along the south coast of the Isle of Wight and between Dover and Folkestone, in each of which cases chalk overlies gault clay. Plate V shows the famous undercliff, west of Ventnor in the Isle of Wight. On the shore between the tide marks the rock fragments are rolled against one another and quickly reduced to rounded boulders, pebbles and sand. These, rolled backwards and forwards between the tide-marks and later dragged below low tide-mark enable the sea to level off its wave-cut platform. This is illustrated in Plate 3. The particularly interesting case of undercutting ot massive limestone shown in Plate XXA is partly due to the small tidal range and the consequent concentration of erosion along one plane.

Thus the effect of the sea round the coasts may be described as the

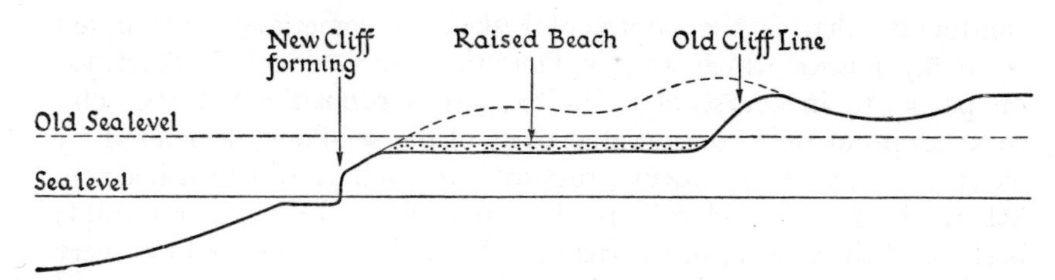

FIG. 19.—Section through a Raised Beach. This is a diagrammatic representation of the scene shown on Plate XI

creation of a platform, a wave-cut rock bench, on which is distributed a veneer of sediments made in the process. The process of its development is shown diagrammatically in Fig. 18. This shelves gently seawards under the water and passes imperceptibly into what is called the Continental Shelf. This is a great shelf found round most of the lands over which the sea is less than 600 feet deep.

The surface of the continental shelf is, normally, very gently undulating through relative resistance of the solid rocks. It is, in fact, a peneplane formed by the work of the sea. Even the slightly irregular denuding action of the sea may result in swellings of the floor which just give rise to shallow areas or may reach the surface as islands. Just as the sea in cutting back a cliff may leave a stack or an island, so in the age-long process of marine peneplanation certain upstanding masses may have been left as islands—it may be isolated and far from land. Where this is the case there is usually an explanation in the hardness or resistance of the rocks of which they are composed. The Scilly Isles are thus the protruding surfaces of an almost submerged granite mass comparable with that of Land's End and from which the surrounding sedimentary rocks have been removed. The isolated mass of Rockall far away in the Atlantic off the north-west coast of Scotland consists of a particularly tough micro-granite and the same is true of Ailsa Craig near the entrance to the Firth of Clyde south of Arran. The celebrated St. Kilda is the largest of a group of sixteen islets rising from the continental shelf. They owe their origin largely to the resistant character of the Tertiary igneous rocks of which they are composed. Three of the small islands of the St. Kilda group are shown in Plate 32A and Stac Lee in Plate XXXII.

It follows that the floor of the "epicontintental" sea around the continents—that is the continental shelf—is sometimes interrupted by rocky masses which are, in fact, the "monadnocks" described on page 214 in course of evolution. It is probable that the solid rocks crop out over considerable parts of the sea-floor, uncovered by sediments, and these "rocky grounds'" are well known to fishermen. Where the rocks are hard and jagged trawling becomes impossible because of the tearing of the nets on the projecting rocks. Over very large areas, however, the continental shelf is covered by a veneer of sediments laid down by the sea itself and derived both from the nearby land by wave action and from the smoothing of the shelf itself as well as from sediments brought down from the heart of the land masses by rivers or ice-sheets. There is normally a gradation from the coarse shingle and stones of the beach near high-water mark, to sand, becoming finer seawards, which in turn passes into silt and mud. This sequence, however, is frequently varied or even reversed: the coastline may yield little or no coarse sediment; rivers may bring down quantities of mud which becomes spread over a wide area; outcrops of rock on the sea floor may yield local spreads of coarse material; and one may get sandy beaches. There are also more local or special conditions which result in variation in the form of the sea floor and its deposits. An interesting case is where icebergs broken off from land ice and bearing a burden of boulders and stones, both on the surface and frozen into the ice, meet warmer water. The icebergs melt and their burden is dropped on the sea floor. This is happening to-day in the Grand Banks area off Newfoundland and it must have happened extensively in the seas round the British Isles during and after the Ice Age. Indeed, boulders floated by ice from distant sources are found in some of the raised beach deposits along the shores of the English Channel.

The variety of habitat for marine bottom-living creatures in the shallow water of the continental shelf, is more than paralleled by the variety of habitat along the sea-shore itself.

Broadly speaking, any stretch of coastline is either one of erosion or of deposition and along such a varied coastline as that of the British Isles the conditions change with great frequency. A cliff coast is obviously an erosion coast and a high or irregular cliff line may be taken as indicative of long-continued erosion. Coast erosion was the subject of an exhaustive inquiry by a Royal Commission which reported in 1911 and much attention was given to the rate of cliff erosion. The

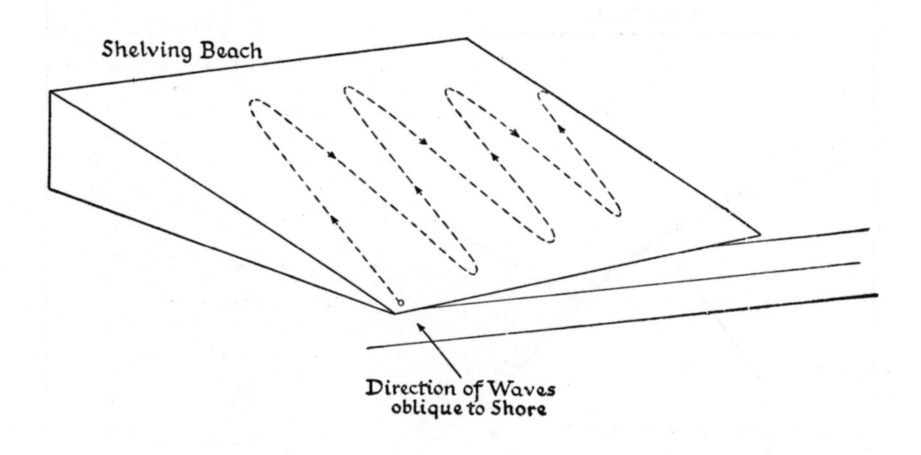

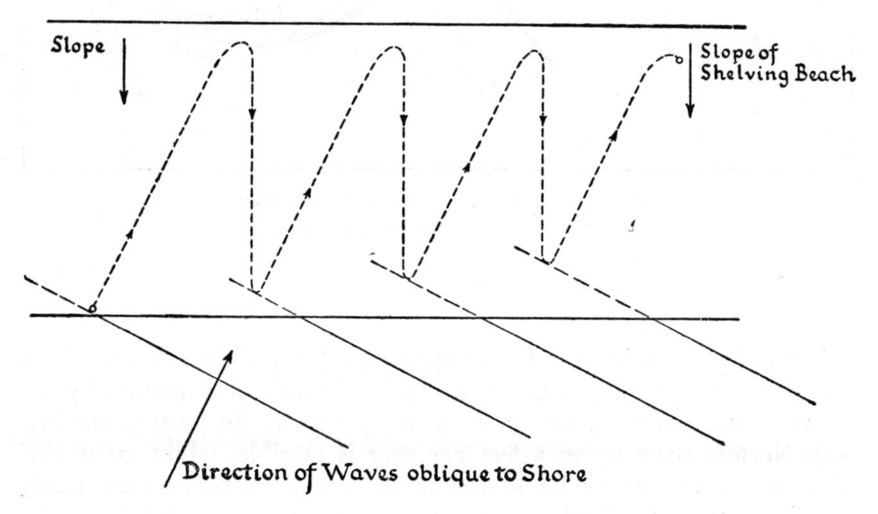

FIG. 20.—Diagrams showing the Drift of Shingle along a Shelving Beach. In each diagram the dotted line shows the course of a single pebble. It is thrown *up* the beach parallel to the direction of the prevailing waves but is dragged *down* the beach roughly parallel to the slope of the beach by the force of gravity. As the process is repeated the pebble works its way, with all its fellows, *along* the shore. If groynes are erected, this longshore drifting is partly arrested (see Plate IX).

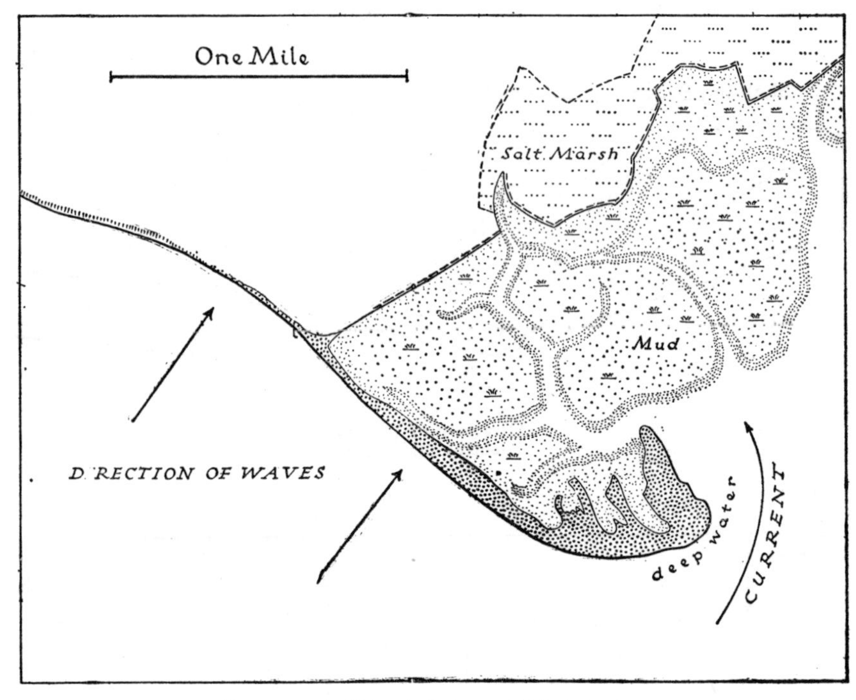

FIG. 21.—A Shingle Spit and Salt-Marshes
Hurst Castle Spit, Hampshire

chalk cliffs of the Strait of Dover appear to be receding at a rate which suggests the widening of the Strait by half a mile in a thousand years —an average of 15 inches on each side per year. In parts of the low soft Norfolk coast up to 5 feet per year is possible, whilst great cliff falls may give spectacular figures in cliffs which otherwise are being eroded at the rate of only an inch or two a year.

A change from erosion to deposition is seen with cliffs in front of which shingle banks are accumulating. Taking Britain as a whole, even without human interference more land is being gained at present than is being lost. The material from a yard or two of high cliff would spread over a large superficial area of mud-flat. Allowing for man's action the gain is very much greater than the loss. Accumulation along a coast may take many forms—among the chief of which are shingle beaches and spits, sandy bays often backed by sand dunes, and mud-flats.

A study of shingle beaches and spits gives a clue to many of the essential features of coastal evolution. For long it was believed, and actively advocated by many writers, that longshore currents were mainly responsible for the drift of shingle and sand along coastlines. That there is a drift of material along shores is quite clear and is immediately apparent where groynes have been built to minimise the movement—the shingle is piled up on one side of the groyne and swept away from the other. This is clearly shown in Plate IX of the shore at Folkestone. In recent years, however, it has been shown, notably by W. V. Lewis, that the drift of material is occasioned by wave action and not by currents. There is a constant tendency for shingle beaches to be piled up so that the ridges are at right angles to the dominant wave direction—in other words the shingle ridges are built up parallel to the ridges and troughs of the waves that create them. Where, owing to the general trend of the coastline, the waves break obliquely, the pebbles are thrown up slightly obliquely but the undertow drags them back more nearly at right angles to the direction of the shore so that they travel gradually along the coast, as shown in Fig. 20. There is, in consequence, the familiar phenomenon of a spit being built out, often right away from the coast line, but at right angles to the direction of the waves. A well-known example is Hurst Castle spit on the Hampshire coast opposite the Isle of Wight. Elsewhere successive shingle beaches may be built up—especially in time of storm, as at Dungeness. Dungeness is remarkable for the extensive area of successive shingle beaches. The spits of shingle or sand which are associated with river mouths are rather different in character : a drift of material deflects the channel of the river. Behind shingle spits and sand spits conditions tend to be favourable for the accumulation of mud and the development of salt marshes.

With sandy shores, wind nearly always takes a hand. The dominant wave direction (as on the coasts of north Cornwall) is often that of the predominant westerly winds. Such areas as Saunton Sands and Perranporth illustrate the formation of sandy beaches at right angles to this direction. Further, these sandy beaches are backed by large areas of sand dunes. Above high-water mark the sand, quickly dried by the wind and sun, is blown inland to form dunes, in due course to be fixed in the usual way by Marram Grass (*Ammophila*) and other sand-dune plants. Wind action is especially important during short or neap tides when an expanse of sea sand below the high-water mark of spring tides is dried and blows easily.

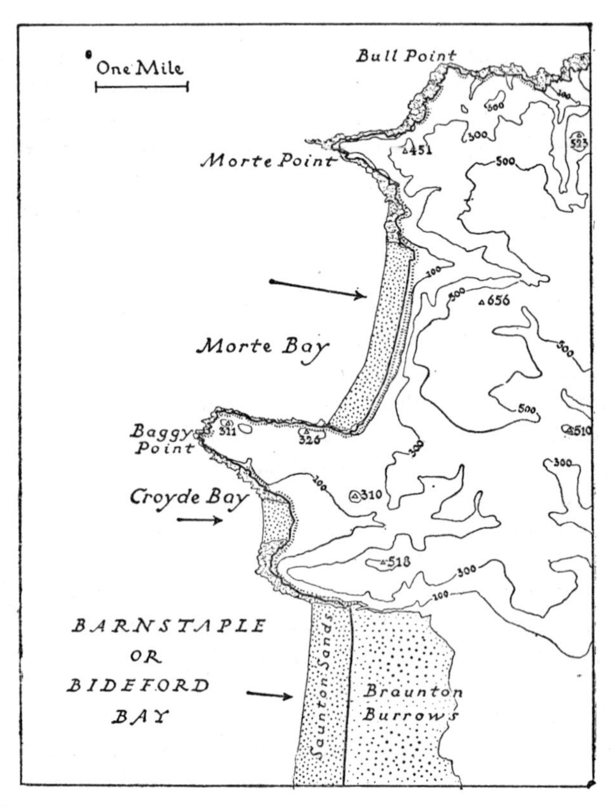

FIG. 22.—An Atlantic Coastline with shingle beaches and sand-dunes at right angles to dominant wave direction (shown by the arrows). Rocky, hilly headlands separate the bays and indicate the east-west strike of the rocks.

Around the shores of Britain there are some remarkably large areas of silt and mud covered only at high tide. Around parts of the Wash high and low tide marks may be several miles apart and the same is true of parts of the Thames estuary, the Bristol Channel, Morecambe Bay and Solway Firth. In these areas there is a steady accretion to the land and the salt-marshes which are developed exhibit the well-known zonation of their vegetation. Reclamation of such areas goes on steadily round many parts of the coast: when the silting has gone on so that the mud-flats are covered only by the higher tides, they are enclosed by earth banks and at first water let in at high tides is allowed to deposit more mud and then to run off gently. Then the entrance of salt water is later prevented and gradually rain water washes the salt out of the soil so that fresh-water marsh replaces salt marsh. After some eight or ten years the ground is sufficiently free from salt for ploughing to be possible.

Such are the features associated respectively with erosion coasts and coasts of accretion. There are other features associated with eustatic movement of elevation or depression. A rising coast frequently shows raised beaches—wave-cut platforms on which rest gravels, sands and other beach- or shallow-water deposits and which are frequently bounded on the landward side by lines of old or "fossil" cliffs. Such raised beaches are well seen round many parts of Britain; especially famous are the examples along the Clyde estuary. Along the rocky coast of Cornwall the raised beaches are but narrow platforms cut in the hard rocks. Of other striking examples from Scotland one has been chosen for illustration in Plate XI.

A sinking coast or drowned coastline is often highly indented for the sea naturally invades the mouths and lower sections of river valleys and runs up the valleys of tributary streams. Excellent examples are seen along the south coasts of Devon and Cornwall, as illustrated in Plate 26. Local features such as submerged forests are indicative of a sinking or a sunken coastline. A fascinating little book by the late Clement Reid on *Submerged Forests* described the very numerous examples round the British Isles. Summer visitors perhaps know best those along the coasts of East Anglia and Lincolnshire, West Cornwall or Cheshire.

A distinction is frequently drawn between the "Atlantic" and "Pacific" types of coastline, so-called because of their relative prevalence round those two oceans respectively. In the Atlantic type the "grain" of the country, the axes of the folds in the rocks, is at right angles to the dominant direction of the coast. This is beautifully

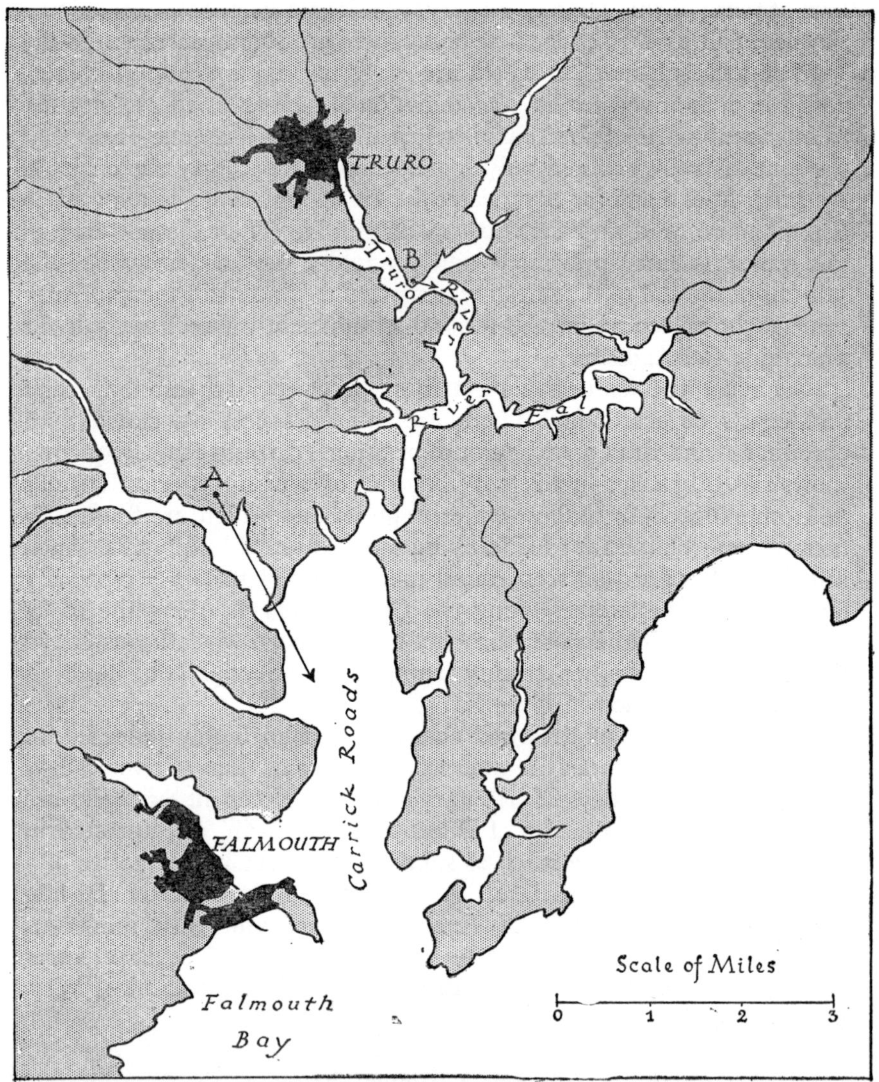

FIG. 23.—A Drowned Coastline

The view shown in Plate 26A is taken from the point A, looking in the direction of the arrow ; the view in Plate 26B is from the point B.

seen in north Cornwall where the Armorican folds have east-west axes whereas the coast often trends more nearly north and south. Banks of resistant rock run out to sea as headlands or ridges of rocks and between them are the secluded bays which form such a delightful feature of the coast (see Plate 3B). The same phenomenon is seen on a larger scale in south-west Ireland, where the hill ridges of Old Red Sandstone are separated by long narrow rias. In the Pacific type of coastline, less commonly seen in Britain, the coast is roughly parallel to the grain of the country. This may be illustrated from the part of the Dorset coast shown in Plate XXVI.

We have referred to the continental shelf and are perhaps little concerned with what happens beyond its edge. The continental slope to the depths of the open ocean are often very steep—so steep that deposits laid down may slide in great masses from the margins of the shelf to the deep ocean floors and give rise to phenomena of extensive slip faulting and puckering which some geologists claim recently to have recognised in the rocks of past geological ages.

Since this book was first published, Professor J. A. Steers has written the volume on *The Sea Coast* for the *New Naturalist* series and reference may also be made to Ian Hepburn's *Flowers of the Coast.*

CHAPTER 7

THE SCENERY OF THE SEDIMENTARY ROCKS

THE GREATER part of Britain is made up of rocks which were laid down under water as sediments. They were originally deposited in almost horizontal layers or strata and when first raised up above the level of the sea they were approximately horizontal. Some have remained so, but for the most part they have been tilted and folded to a greater or lesser extent. Interbedded with the sedimentary rocks proper are limestones, as well as lava flows and beds of volcanic ash. Whilst some limestones are built up of fragments of shells and are thus sediments, others are of the nature of chemical precipitates and again others are coral reefs which have grown in situ. Hence the distinction from sediments proper.

Whilst some of the members of the sequence retain the same general characters through a great thickness—the Chalk, for example, in places is nearly 1000 feet thick—it is a common feature for strata of different types—clays, shales, sandstones and many others—to succeed one another with almost bewildering rapidity so that each bed may be only a few feet, or even a few inches, in thickness. This is of great importance since each bed may, and usually does, differ in its resistance to atmospheric and other forms of weathering. It is this variation in ability to stand up against the denuding forces of nature which is largely responsible for the intimately varied scenery of so much of Britain and the absence of monotony even in the lowlands. It is also largely responsible for the remarkable and rapid variations in the sea coasts of the British Isles.

Before considering the land forms which are characteristic of the sedimentary rocks it will be well to recall a few of the salient features of the rocks themselves. The majority show evidence of bedding planes, indicating that they were deposited in layers. This is generally true of the harder rocks such as the sandstones, but bedding planes may be absent in the clays which form an amorphous mass. A few sands and sandstones which were deposited in shallow waters under the influence of currents have a curious type of bedding consequent upon current action and hence known as current bedding or false bedding. It is best understood by reference to the picture, Plate 10 and the diagram Fig. 37.

The bedding planes tend to form planes of weakness along which the rocks split easily. When a hard sandstone splits easily in this way it is known as a flagstone. Because of this property the Caithness flagstones of Old Red Sandstone age were the favourite paving stones for London and other cities till the introduction of artificial stone for the purpose. Some rocks break into such thin slabs that they can be used for roofing and though true slates have a different origin (see p. 81) such "slates" as those of Stonesfield and Collyweston are actually thinly bedded limestones. The vogue for "crazy paving" in gardens has introduced many townsmen to a considerable variety of rocks which split easily along the bedding planes. Rocks which do not split more easily along the bedding planes than in other directions are referred to by masons as "freestones" because they are freely worked into building blocks. Even so, as the expert knows, if a freestone block is laid on its edge it is liable to weather more rapidly than if placed as it was quarried. The term freestone is applied both to sandstones and to limestones.

In addition to this character of planes of weakness parallel to the bedding, the majority of the harder rocks have cracks or planes of weakness at right angles to the bedding along which they split easily or along which they are easily attacked by denuding agencies. These planes are known as joints or joint planes and there may be more than one series. In the latter case the major series is known as that of the master joints. When a rock such as a hard sandstone is well jointed it often weathers into vertical crags—a feature exhibited to perfection by the Torridonian Sandstones of the north-west of Scotland (see Plate 1B). The craggy sandstone cliffs of such heights as Stac Polly afford interesting plant and animal habitats. Although the jointing in limestone is partly of a different origin, being due to some extent to solution of the rock, the effect is broadly the same (see page 74 and Plate 6). Very naturally jointing and bedding both influence to a great extent the character of sea-cliffs and river cliffs. A well-developed system of joints results in vertical cliffs; where there is at the same time a dip of the rocks seawards or riverwards the combination of circumstances is such as to encourage the detachments of great blocks and the creation of landslides (see page 192).

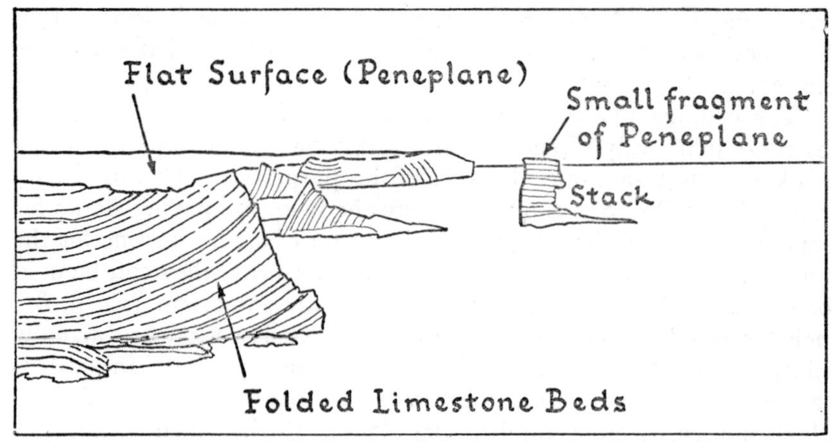

FIG 24.—Diagrammatic Explanation of the Peneplanation shown in Plate 5A
A typical example of submarine peneplanation, with a subsequent uplift of the whole surface. As a result the almost level surface of the ground is independant of the underlying geological structure (see also p. 219 and p. 221).

Sedimentary rocks when laid down are almost horizontal so that the simplest case to consider is when horizontal strata are raised up and subjected to denudation. Rivers cut into them and between the young

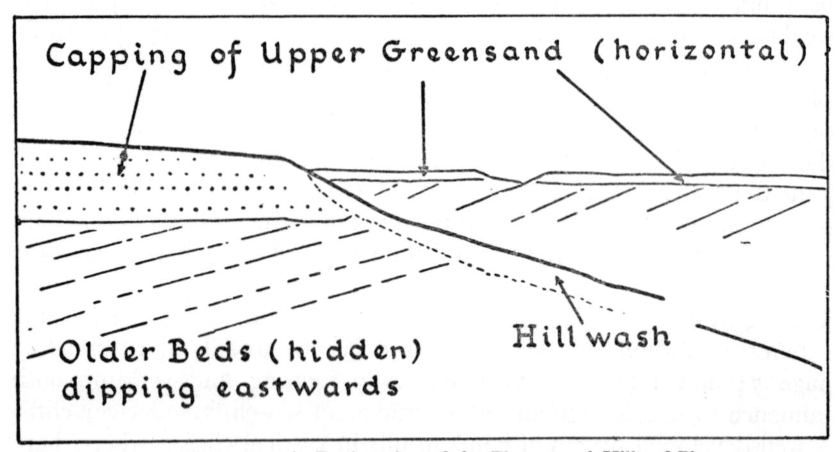

FIG. 25.—Diagrammatic Explanation of the Flat-topped Hills of Plate 5B
In contrast to Fig, 24, the flat tops of the hills here follow the bedding of the underlying Upper Greensand which rests almost horizontally, but unconformably, on underlying rocks which dip to the east (see also Fig. 27).

L. D. STAMP

The coast of South Pembrokeshire
The flat surface is due to marine peneplanation of folded Carboniferous Limestone

L. D. STAMP

Flat-topped hills near Honiton, Devon
The flat surface is due to the horizontal capping of Upper Greensand

PLATE 6

L. D. STAMP

Cheddar Gorge, Somerset
A typical dry limestone gorge (See page 74)

streams are flat-topped plateaus coinciding with the horizontal beds. The gravel terraces of a river are really of this nature—a horizontal bed of gravel has been deposited by the river which has then, by the rejuvenation described above on page 48, been cut into by the river. Sometimes gravel is sufficiently bound together by "hard pan" or "iron pan"—ferruginous cementing deposits—to form a "cap" resistant to weathering and so protecting softer rocks below. Thus Shooter's Hill near Woolwich consists of London Clay but has been protected from erosion by a capping of ferruginous gravel. On a somewhat larger scale a hard or resistant bed in a series of mixed strata normally forms the capping of flat-topped "mesas" and "buttes." An interesting feature results when a conical hill is capped by the remains of a once extensive stratum of hard rock. Such hills are well known in the more arid parts of South Africa under the name of kop (plural kopje) but hills of similar origin though with sides more smoothed and vegetation-covered as a result of our wetter climate may be seen near those parts of the Cotswold edge where the beds are almost horizontal. Bredon Hill, Worcestershire, is a good example. This hill might be called a butte because it is isolated from the main plateau. Seen in plan these flat-topped plateaus are usually very irregular and are cut into by valleys which are often deep, narrow, steep-sided and winding. The Cotswold valleys in the Stroud area are good examples.

The position is changed when the beds, instead of being horizontal, are tilted or dipping. Then the hills to which the more resistant rocks give rise, instead of being flat-topped plateaus, present a long gentle slope on one side—the dip-slope, corresponding broadly with the dip of the underlying rocks—and a steep slope, the scarp slope, on the other. In plan the crest line of the line of hills tends to form a relatively straight line or is arranged in long sweeping curves. In the south and east of England, where the whole sequence of Jurassic and Cretaceous rocks consists of an alternating series of weak clays and sands with resistant limestone and sandstones, these lines of hills with their long, almost imperceptible dip slopes and their abrupt scarp slopes, from the crest of which so many magnificent sweeping views may be obtained, are the dominant feature of the landscape. The rocks dip generally to the east, south-east and south so that the scarps face to the west (as in the well-known and remarkably straight Lincoln Cliff), to the north-west (as with the Cotswold and Chiltern scarps) or to the north (as with the Blackdown Hills of the Dorset-Somerset

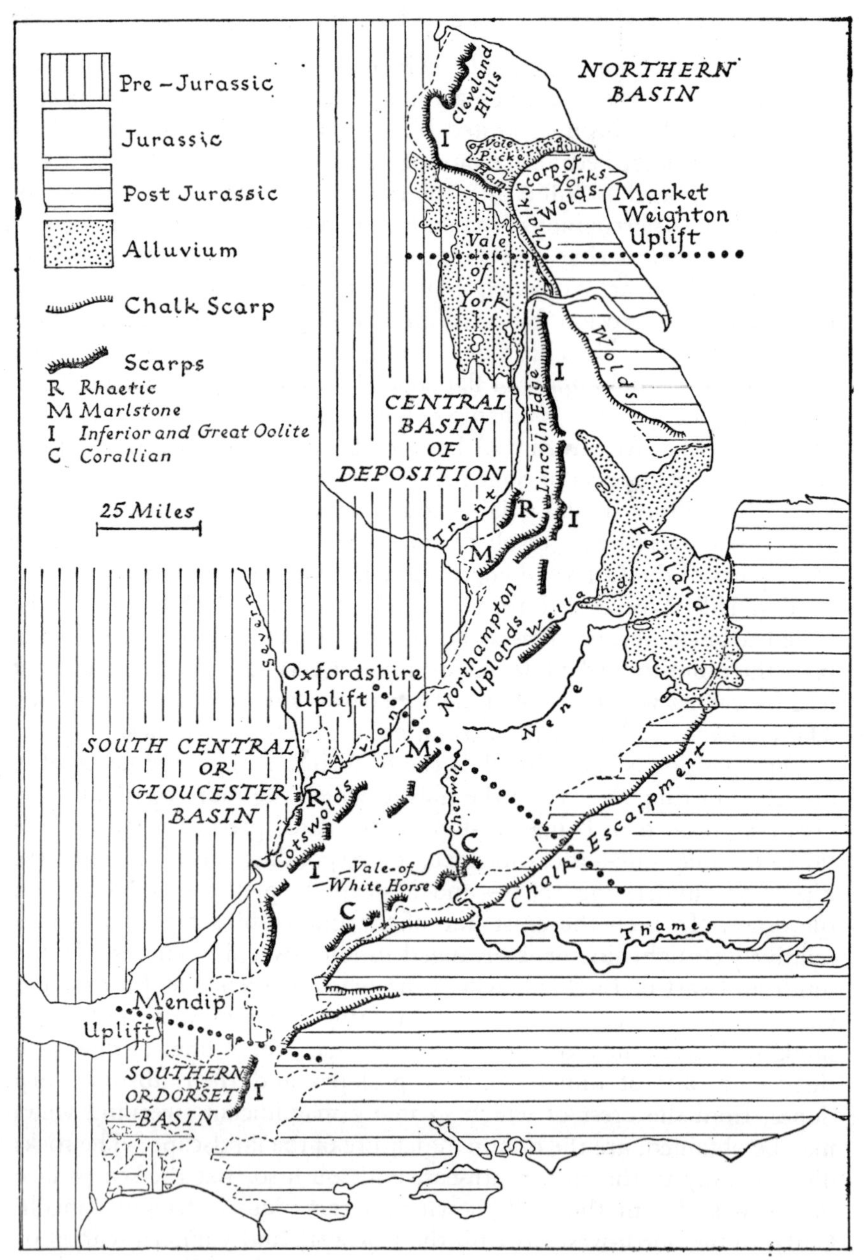

FIG. 26.—The Scarplands of England, showing all areas where the scarp faces have a gradient of more than 1 in 7 (after S. H. Beaver)

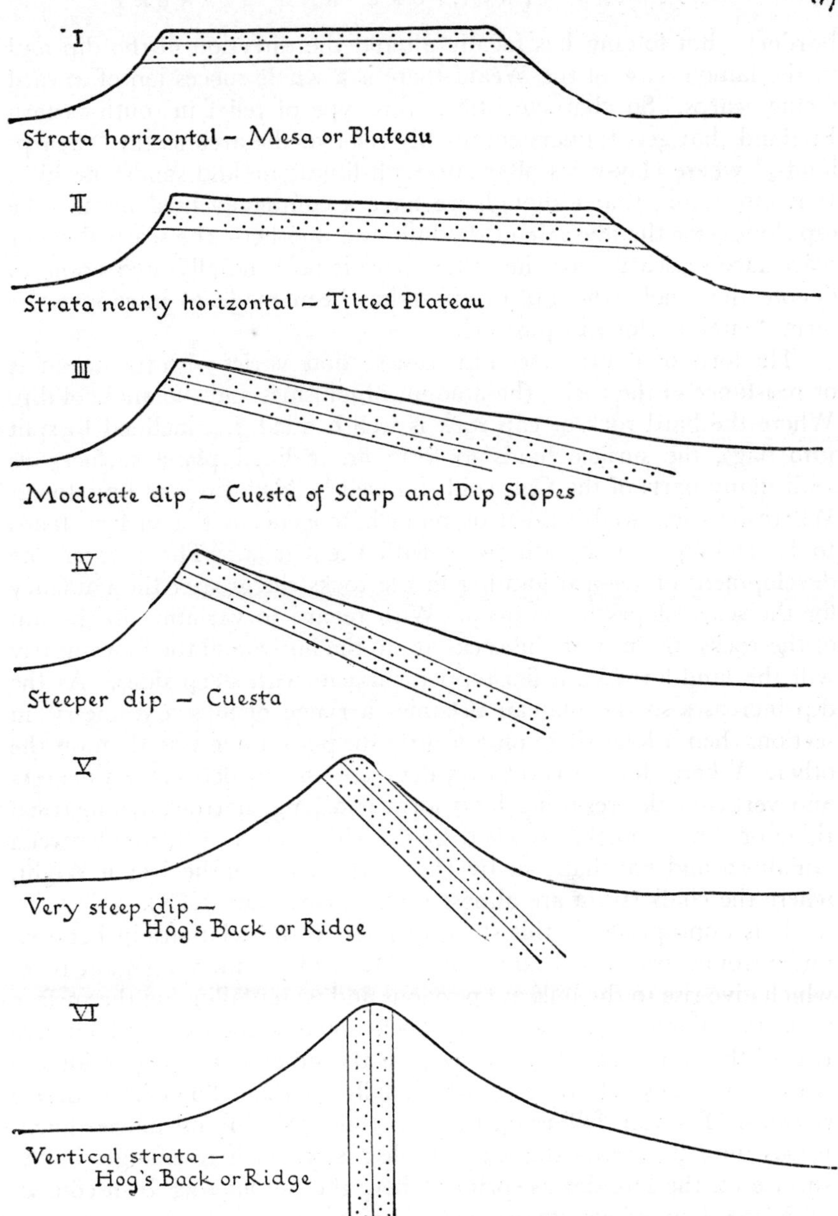

FIG. 27.—Forms of Cuestas, showing relation of surface form to dip

border). But folding has locally altered the direction of the dip and in the famous case of the Weald there is a whole succession of inward facing scarps. So characteristic is this type of relief in south-eastern England that geographers commonly refer to the area as the " Scarp-lands," where clay-vales alternate with limestone and sandstone hills. It is interesting that although we have a commonly used name—the dip slope—for the one side of the hills and another—the scarp slope or scarp face or scarp—for the other there is no generally-used name to denote the whole, though an effort has been made to introduce the term " cuesta " for this purpose.

The form of a cuesta, seen in cross section, varies with the hardness or resistance of the rocks, the amount of jointing, and the angle of dip. Where the hard rock or cap rock is well-bedded, i.e. inclined to split into flags, the surface tends to form an inclined plane surface—as with many parts of the Cotswolds on well-bedded Jurassic limestones. Where it is less well-bedded or more homogeneous the surface tends to be rolling or undulating—as with the Chalk. The greater the development of vertical jointing in the rocks the greater the tendency for the scarp slopes to be steep. With regard to variation in the dip of the rocks, the nearer the rocks are to the horizontal the more nearly will the land-form be a flat-topped plateau with steep sides. As the dip increases so the plateau becomes a range of hills asymmetric in section, then a long ridge only a little steeper on one side than on the other. Where the dip is over 45 degrees—that is between 45 degrees and vertical—the resulting land form is a long, narrow, symmetrical ridge or " hog's back," typified by the Hog's Back of Surrey between Guildford and Farnham, or by the central ridge of the Isle of Wight, where the chalk strata are almost vertical (see Plate XIX).

It is appropriate at this stage to refer to the relationship between underground water and land forms. Most of the more resistant rocks which give rise to the hills are pervious and water falling on the surface soaks into the ground. The weak rocks, such as the clays, which give rise to the intervening valleys are, on the other hand, impervious to water, and most of the rain which falls on them flows into surface streams. The rain falling on the hills thus soaks in and moves downwards until it reaches the water table ; some of it re-emerges at the surface on the hillsides as springs where the porous rock is in contact with the impervious underlying clays.

We turn now to the consideration of land forms which are developed

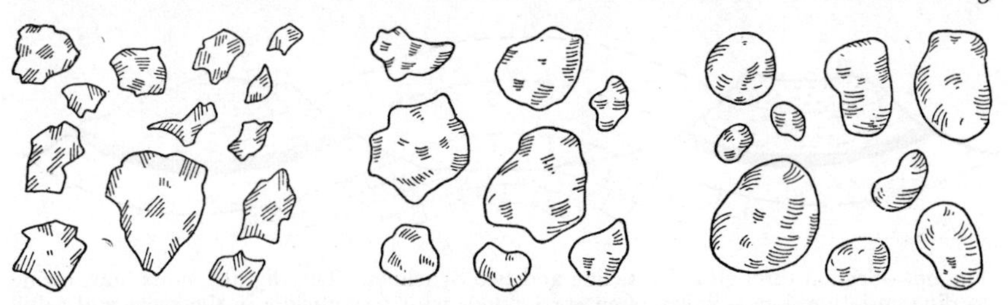

FIG. 28.—Sand grains under the microscope. Those in the centre are typical of sea-sand, approximately rounded but prevented by the "cushioning" effect of the water from becoming exact spheres. Those on the right are almost perfectly rounded and polished and are typical of wind-blown sand found under desert conditions. The angular grains on the left have simply been crushed as happens when a great mass of ice grinds rocks one against the other.

where the rocks are folded. Where the rocks are gently folded it might, very naturally, be thought that an upfold or "anticline" would give rise to a hill range and a downfold or "syncline" to a major valley. When the folds are first formed this is probably the case, but such an apparently simple relationship is comparatively rare in nature. Where the upfold is of a large mass of resistant rock of homogeneous composition it may remain as such, but where as is more usually the case the upfold consists of a varied series of resistant and "weak" strata, the forces of denudation get to work and, once they have cut through the protecting arch of resistant rock, proceed to develop what are in fact series of scarp and dip slopes. Any softer rocks in the centre of the anticline, robbed of their protective capping, are rapidly excavated and so it is very usual to find the main valleys corresponding with the axes of anticlines whilst the intervening hill ranges are synclinal in structure. This is well seen in many parts of north Wales—Snowdon itself is synclinal in structure (see Plate 29). The relationship is clear from the photograph.

So far we have referred to simple structures due to one major resistant rock overlying a less resistant. What happens when there is a rapid alternation? The principles are the same—a succession of small cuestas may result, or very commonly a "stepped" scarp. Thus the foot of the Chalk scarp has often a platform built up on the Lower Chalk, at the top of which is a relatively hard bed: at the foot of the main Cotswold scarp there is often a similar platform. The asymmetry and shift of valleys (see Fig. 30) is similarly related to structure.

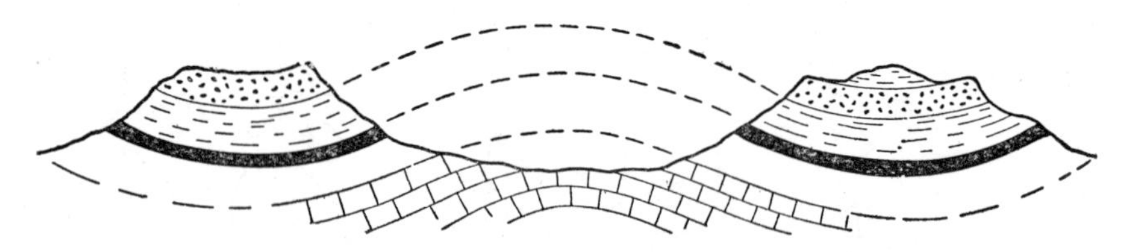

FIG. 29.—Section through an Anticline and two Synclines. This digram shows how, owing to differential erosion, a valley often corresponds with an anticline in the rocks and a hill or mountain with a syncline. A careful study of Plate 29 shows that Snowdon is synclinal in structure—not unlike the hill shown on the right of this diagram.

Some resistant beds, for example the great Chalk formation, stretch uninterruptedly for great distances and give rise to features which are correspondingly constant. In other cases the resistant beds may be of relatively local occurrence and the cuestas associated with them may die away. This is true of the Jurassic limestones and there is no one continuous Jurassic limestone scarp across England as there is with the Chalk. Instead, there are scarps of more local occurrence associated with individual beds as is indicated in Fig. 26.

It is important to remember that the land forms described in this chapter as associated with sedimentary rocks belong to a comparatively youthful stage in the evolution of a land surface. As the cycle of erosion progresses and sub-aerial denudation tends towards the peneplanation of the whole surface (see pages 40-41) features which are due to varying resistance of the rocks and to varying structure gradually become eliminated as the surface is reduced to a monotonous level or almost level plain. Similarly if the land sinks below the sea and is subjected to sub-marine peneplanation and then subsequently raised again to form a land surface, the hills and valleys will have been eliminated and the monotonous surface will bear little or no relationship to the underlying structure. Both sub-aerial and sub-marine peneplanation have happened repeatedly in the long geological history of the British Isles so that in many areas the land forms are due primarily to other causes than to underlying structure. Then it must not be forgotten that all Britain north of the Thames-Severn line was covered by ice-sheets and that glacial action has profoundly modified the pre-existing land-forms.

The closest relationship between the structure of the sedimentary rocks and the present land surface is thus to be found in the south-east

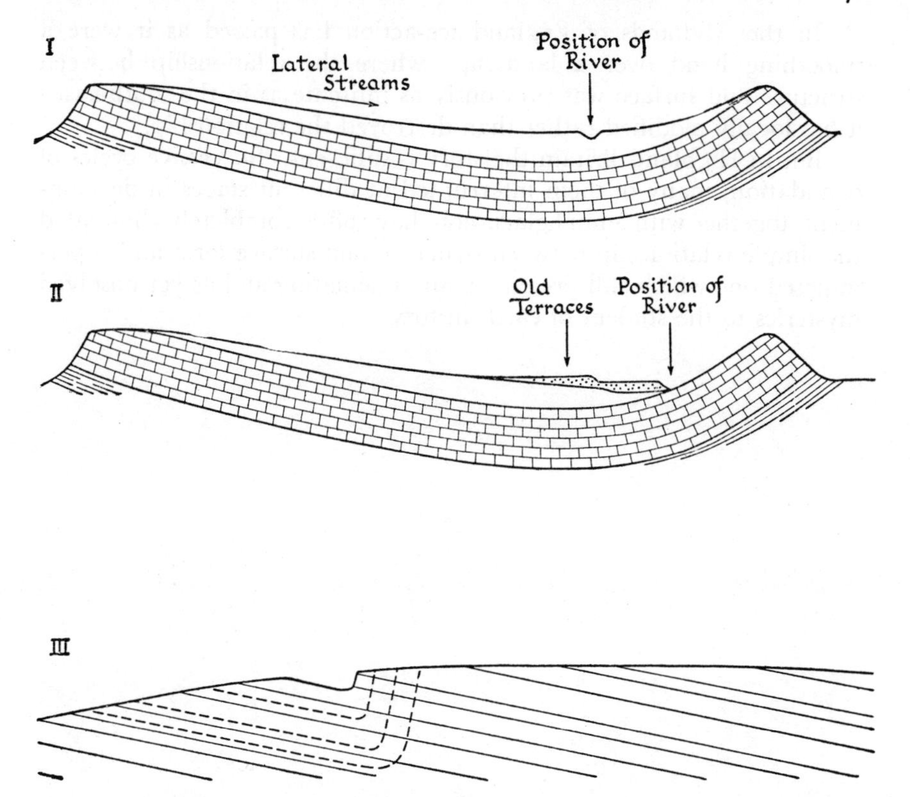

FIG. 30.—Sections showing the effects of an asymmetric syncline (I and II) or dipping strata (III) on the shift and form of a valley. The upper two diagrams might represent the Thames near London with the Chiltern Hills some distance to the north and the North Downs considerably nearer on the south. They explain why the Thames terraces (II) are more extensive in Middlesex and Essex, north of the River, than in Surrey or Kent.

and south of England where Alpine earth-movements caused folding and the geological time which has elapsed has been insufficient to permit peneplanation where ice-action has been absent. Even here rises and falls in relative levels of land and sea have produced complications due to marine denudation and complex river action. In the Weald the broad features are clearly dependent on the major structure of the area : detailed examination shows minor features of the greatest interest such as wave-cut benches (see p. 52) revealing a long and complicated geological history.

In the Midlands of England ice-action has passed as it were a smoothing hand over a landscape where the relationship between structure and surface was previously as intimate as in the south-east; it has locally modified rather than destroyed the relationship.

But in Highland Britain the story is different. Successive cycles of denudation, many of them interrupted at different stages in development, together with a final glaciation, have often completely eliminated the simple relationship between structure and surface form and superimposed one which still presents many fascinating and as yet unsolved mysteries to the student of earth history.

THE SCENERY OF LIMESTONE COUNTRY

LIMESTONE is different from all other rocks of widespread occurrence in this country in that it is soluble in weak acid—even including rain water which has absorbed carbon dioxide from the atmosphere. With other rocks, rain water which soaks into the ground finds its way slowly or quickly according to the permeability of the rocks to the permanent water table and it behaves according to well-defined laws ; it may wash out the fine particles but does not dissolve the rocks on the way. In limestone country, on the other hand, the rain water which sinks into the ground finds its way into fissures in the rock and there slowly and surely, though unseen and often unsuspected at the surface, it proceeds to dissolve away part of the rock, forming underground watercourses and caverns. When the water emerges as springs it is highly charged with calcium bicarbonate and though a limestone may be a good reservoir of water and yield an excellent supply the water will be hard—" temporary hardness." This " hardness " is eliminated by boiling, when the carbon dioxide is driven off from the soluble calcium bicarbonate and calcium carbonate or limestone is redeposited.[1] Permanent hardness, due to the presence of calcium sulphate, is not eliminated by boiling.

Once an underground channel has been formed, even if it is no more than the widening of rock fissures, rain falling at the surface tends at once to drain underground into the channels so that in limestone country surface streams are few. Even with large streams the water not infrequently disappears into " swallow holes " perhaps to reappear from similar holes lower down the valley or from the mouth of a cave.

Once the underground caves have been formed a whole cycle of solution and redeposition of the limestone may occur. Drops of water percolate through the roof of the cave, and may remain there long enough for the water to evaporate and leave behind a thin film of crystalline limestone. The process is closely comparable with the formation of icicles—and the hanging " icicles " of calcium carbonate are called stalactites. If the drops of water fall to the floor and there

[1]According to the simple formula $Ca\,H_2\,(CO_3)_2 = Ca\,CO_3 + H_2O + CO_2$

evaporate the deposit left behind forms stalagmite as opposed to stalactite on the roof. It may form an even layer or may be built upwards into all the fantastic forms so well known from the limestone caverns of the country. A good example of a subterranean stream passing through a succession of caverns and then emerging from the mouth of one is afforded by Wookey Hole in the Mendips of Somerset. Nearby is the vertical-sided Cheddar Gorge—the supreme example of a limestone gorge which may have resulted from the roof of the line of caverns collapsing and subsequent torrent action (see Plate 6).

Limestones of three main types are of widespread occurrence in Britain—the older, crystalline limestones of which the chief is the Carboniferous Limestone ; the bedded limestones of the Jurassic system, with which may also be grouped the Permian Magnesian Limestone ; and the Chalk. Each gives rise to characteristic land forms.

The hard, crystalline Carboniferous Limestone forms the Mendip Hills, part of the rim of the South Wales Coalfield, as well as the Gower Peninsula, considerable areas of the Pennine country of Derbyshire and of the Pennines farther north, and an intermittent ring of country round the Lake District. The Devonian limestones of the Plymouth and Torquay neighbourhoods are similar and so are the Silurian limestones of Wenlock Edge and other areas of the Welsh Borderland and of the Wren's Nest near Dudley. All these are literally hard rocks and like other hard rocks of the older geological periods have been extensively fissured. They tend on the whole to form upland country and outcrops of bare rock are frequent. The weathering of the crags can produce charming colour effects with the pale grey of the limestone contrasting with the greens of the vegetation. Considerable bare areas (limestone pavements) are a common feature of parts of the Pennines and these may be seamed with deep solution grooves, termed "clints" or "grikes." Because of the power of water to dissolve the limestone, rivers tend to eat their way downwards between almost vertical walls and limestone gorges are a feature of the larger rivers—for example of the Dove and Derwent in Derbyshire and the gorge known as the Winnatts above Castleton. A similar example is shown in Plate XIV.

The phenomena associated with limestone country are well displayed in the famous "karst" area of the Istrian peninsula and northern Yugoslavia and in consequence it is usual to talk of karst country and karst phenomena. The British limestone country differs in its well-distributed rainfall and the consequent well-developed vegetation cover from the arid karst lands proper. The term karst is thus of limited application to Britain, though some geologists and geographers refer to all the features of limestone country as karst features.

If the limestone is impure—as most limestones are—an insoluble residue is left when the lime is all dissolved. Nodules and bands of chert commonly occur in the older limestone and are left when the limestone weathers, but interest centres on the clay which collects in hollows in limestone country and represents the normal insoluble residue of an impure limestone. It is often iron-stained and red ; the lime is completely leached out and so in most limestone country there are pockets of clay giving rise to acid, heavy soils and supporting a calciphobous vegetation in complete contrast to that on the neighbouring limestone outcrops.

The Jurassic limestones vary greatly in character and purity. They are all bedded limestones and resistant to weathering in comparison with associated clays so that they stand up as scarps—well seen in the Cotswolds. A good example of Cotswold scarp scenery is shown in Plate 24. Unlike the older limestones, rock outcrops are few, for the rocks are usually covered by a thin brownish or reddish soil, in large part composed of the insoluble clay residue from the limestone. The rocks are both fissured and bedded so that underground water finds its way through the network of fissures without the need of excavating larger underground channels, and thus caves are relatively rare. At the same time the rocks are harder than the chalk—they afford some of England's finest building stones such as the Bath and Portland Oolites and the many types of Cotswold stone.

The rolling downland of the Chalk lands is one of the most distinctive types of scenery in Britain. Though resistant to weathering compared with the Gault clay which lies below it, chalk is a soft white limestone—pure and dazzlingly white in the upper part, becoming grey, marly and less pure in the lower parts. In the Upper Chalk are many nodules and broken layers of flints, white on the outside, black and lustrous when broken ; in the Middle and Lower Chalk flints are few. Flints are a form of silica which is soluble to a certain extent in water and are due to the redeposition from solution of silica originally scattered through the chalk as sponge spicules and other remains of siliceous organisms. Flint was sometimes redeposited from solution along bedding planes and the tabular flints so formed were particularly valued by prehistoric man for the manufacture of implements. Incidentally the making of flints for flint-lock weapons is an industry which still survives at Brandon in Suffolk.

Chalk is much fissured and rainwater easily finds its way without the solution of underground channels which are rare, but " piping " or irregular solution of the surface is common. The insoluble residue from the solution of the limestone is a brown clay full of angular flints —the well-known Clay-with-Flints. Plateau surfaces are thus often covered with a deposit of Clay-with-Flints from a few inches to many feet in thickness and the same deposit fills the " pipes." It gives rise to very heavy acid soils in complete contrast to those derived from the chalk itself. Some typical pipes in the chalk near Canterbury filled with Clay-with-Flints are shown in Plate XIII.

The most distinctive features of chalk scenery are the numerous branching dry valleys. They are arranged as are the tributary valleys of rivers yet have no streams. They may have been excavated partly by ice sludge (see page 89) but would seem to owe their origin mainly to river action when the water table was higher, before and soon after the Ice Age. An example is shown on Plate 23. Not infrequently the scarp edge of the chalk is notched by dry gaps or wind gaps which are really dry valleys beheaded by the recession of the scarp. Natural rock exposures (apart from sea cliffs) are almost unknown but there are few areas of chalk where there are no quarries to show the underlying rock.

CHAPTER 9

THE LAND FORMS OF VOLCANIC COUNTRY

SINCE there are no active volcanoes in Britain, to devote a chapter to the land forms and scenery of volcanic country may seem unnecessary. Apart, however, from widespread volcanic activity associated with the Archean, older Palaeozic, Devonian, Carboniferous and other periods, and the major role played by the intrusion of great masses of granite and other rocks, since exposed by denudation, in the great mountain-building periods, the Tertiary was marked in parts of the British Isles by widespread volcanic activity. Under the stress of the Alpine movements which folded some of the younger rocks in the south of England, the older rigid blocks of the Highlands were severely cracked and it was through these cracks that great masses of molten lava reached the surface. There are, accordingly, many parts of Highland Britain where igneous rocks of varied types play the chief part in determining the land-forms and scenery. In terms of geological time some of these rocks, dating from the middle Tertiary—especially Miocene—are recent compared with the ancient rocks by which they are surrounded, with the result that the associated land forms are clear-cut examples of those due to volcanic activity.

Popular interest and imagination couples volcanic activity with the cone-shaped volcanoes of the type of Etna or Vesuvius—built up of successive layers of ash blown out of the central vent or neck, the outbursts of volcanic ash being interspersed with outpourings of lava. A final stage is often reached when lava fails to reach the surface and solidifies in the neck of the volcano, effectively choking it. The lava masses are usually far more resistant to weathering than the ash beds so that when the surrounding ashes have been completely worn away the volcanic necks remain as isolated conical hills—often very steep sided. Hills of this origin are a feature of the scenery of much of the central lowlands of Scotland—Castle Rock on which Edinburgh Castle stands, the Bass Rock, Inchcape Rock, Traprain Law, and North Berwick Law are all examples. Arthur's Seat, Edinburgh, shown in Plate XVIA, is a complex example with two vents. All these are of the Carboniferous age.

Sometimes, though not very frequently, it is possible to connect the remains of lava flows, interbedded with sedimentary rocks, with the vents from which the lava emerged.

Though some of the Tertiary volcanoes of western Scotland were of the neck and crater type, for the most part the lava simply welled out of great cracks or fissures. The lavas were mainly basalts—basic in composition and very fluid when molten—so that the lava spread evenly over very large tracts of country, in contrast to acid lavas which are very sticky when molten and do not flow far or spread evenly. One

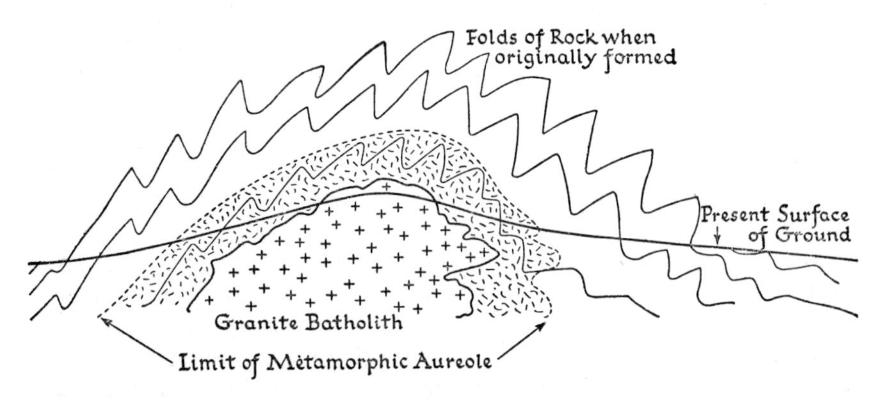

Fig. 31.—Section through a Batholith, showing the metamorphic aureole. This diagram shows the general structure of one of the granite masses of Devon or Cornwall (from South to North) and why the metamorphic aureole is there wider to the south of each mass than to the north.

typical feature is thus the lava plateau. The marked tendency of basalt when cooling to form six-sided columns at right angles to the surfaces is well known at Giant's Causeway and the Isle of Staffa (shown in Plate XVA). As a result of such a perfect system of vertical joints the edges of lava plateaus often stand out as black, forbidding, vertical cliffs, such as those overlooking the glens of Antrim. Similar dark basaltic cliffs stand out amongst the light coloured Carboniferous Limestone of the Pennines as at High Cup Nick. The largest of the lava plateaus in the British Isles is that which covers 1600 square miles in Antrim. It has sagged or collapsed in the middle and is there occupied by the shallow waters of Lough Neagh which though 153 square miles in area is nowhere more than 56 feet deep. Another very large lava plateau occupies the northern part of the island of Skye; a third makes up much of the Isle of Mull.

In many areas the Alpine movements simply shattered the rigid Archean rocks so that they became seamed with innumerable parallel cracks. Into these cracks molten rock forced its way so that dykes were formed, so numerous that the expression "dyke swarm" is often used. Where the basalt of the dykes is much harder than the surrounding rocks the dykes stand out as wall-like masses and they are often conspicuous on coasts. Amongst well-bedded rocks the lava often forces its way more or less horizontally between the bedding planes as a sill. A sill acts usually as a resistant member in the sedimentary

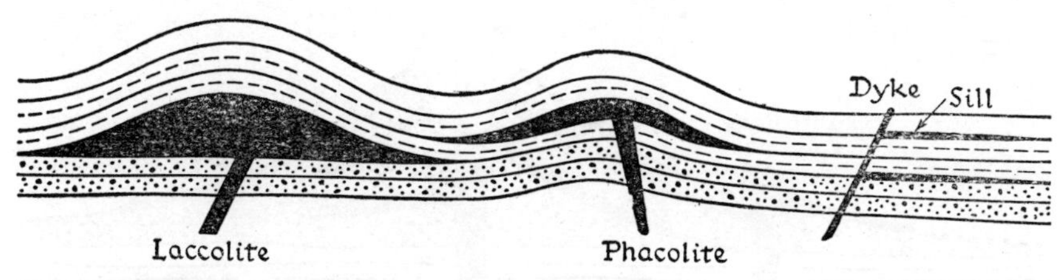

FIG. 32.—Section through a Laccolite and a Phacolite and through two sills and a dyke. In each case the igneous rock, which has been forced up in a molten condition from the lower layers of the crust, is shown in black.

sequence—giving rise to steep-faced escarpments. Notable examples include the famous Whin Sill of Northumberland and the Romans, recognising the natural obstacle afforded by the steep face of the sill, built their defensive wall (Hadrian's Wall) to reinforce it. Another interesting though small example of a sill is Salisbury Craigs, Edinburgh, shown in Plate XVIA.

Lavas interbedded with sedimentary rocks and often alternating with beds of volcanic ash are particularly common in the Ordovician rocks, notably in Wales, and occur also in the upper part of the Cambrian and the lower part of the Silurian. Sills are also found and it is frequently difficult to differentiate between sills and interbedded lavas. Both are resistant rocks and play a leading part in determining the scenery. Most of the higher and more rugged mountains of Wales are built up largely of igneous rocks—notably Cader Idris and Snowdon.

A special type of sill is one where the lava swells out to form a lens-shaped mass—which according to its particular form is known as a laccolite (with a flat base) or a phacolite (with a curved base). On weathering these may give rise to dome-shaped hills—one of the best British examples is the Corndon in south-west Shropshire.

The deep-seated movements of molten rock which are associated with the great mountain-building periods result in the intrusion towards the surface of huge masses of magma, which later, since they are still covered by thick layers of the crust, cool slowly, with the result that the constituent minerals have time to assume a crystalline form easily visible to the naked eye. These masses naturally bake and harden the rocks with which they come in contact, forming the metamorphic

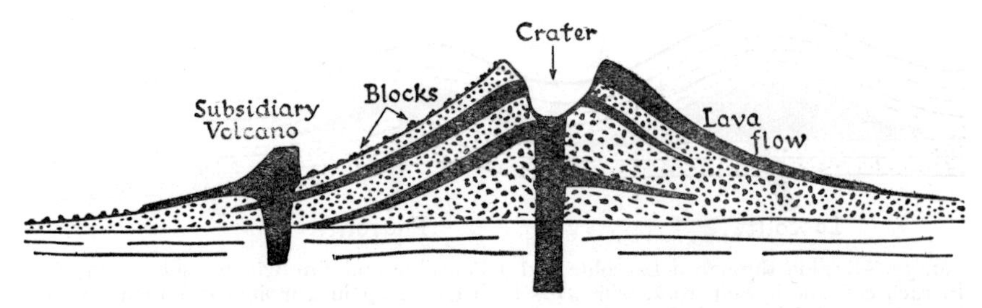

FIG. 33.—Section through a Volcano. The form of the subsidiary volcano, with its neck plugged with hard volcanic rock, may be compared with the hill shown in Plate XVB or with almost any one of the isolated " laws " of the central lowlands of Scotland—such as the Edinburgh Castle rock or Stirling Castle rock.

aureole. The association of metalliferous deposits with the metamorphic aureole has already been mentioned (see p. 25). When the covering rocks have been removed by denudation, the great igneous bosses or batholiths of plutonic rock, of which granite is the typical member, resist further weathering and so stand up as hill masses. With the exception of Exmoor, all the great moorland areas of the south-western peninsula are caused by the Armorican (p. 18) granite masses—Dartmoor, Bodmin Moor, the Land's End mass and the Scilly Isles are examples. Granitic masses of various ages are widespread in Scotland and make up such typical mountains as the Cairngorms and Goat Fell in Arran. Some of the masses may be as late in age as Tertiary—the Mourne Mountains of Ireland certainly are.

ANNE JACKSON

Ulpha Fell, Lake District
Looking southwards to Carboniferous Limestone hills (See page 74 and page 223)

B. GARTH

Chalk cliffs, Bempton, Yorkshire (See Fig. 18 ii)

Llyn Du'r-arddu, Snowdonia M. WIGHT
A typical corrie lake to the north-west of Snowdon, 1900 feet

L. D. STAMP
A coastal scree of sandstone and shale, Cornwall, showing angle of rest (See page 46)

Over Scotland the surface of the great granite masses has been smoothed by ice action ; whilst in the south-western peninsula one can see the effects when well-jointed granite is attacked by the sea. The great boulders of the Land's End cliffs, and the interesting tors of Dartmoor and Bodmin Moor (which may have formed stacks or islands when the sea smoothed the surrounding land) are typical examples of land forms resulting from the weathering of granite. An example from Dartmoor is shown in Plate XXVII.

In addition to the changes induced in rocks as a result of their contact with masses of molten magma, rocks may be profoundly altered by intense pressure and folding. The results of this dynamo-metamorphism are often indistinguishable from those of contact metamorphism. In both cases new minerals may be formed and the rocks assume a crystalline character. Where small flakes of mica are formed in fine-grained rocks such as shales these tend to arrange themselves at right-angles to the pressure and cause the hardened rock to split or cleave easily in a direction unrelated to the original bedding. The existence of this " slatey-cleavage " is the distinguishing mark of true slates.

THE SCENERY OF GLACIATION

AT THE period of their greatest extent, ice-sheets covered practically the whole of Britain north of the latitude of the Bristol Channel and the lower Thames. Since the Great Ice Age is comparatively recent in terms of geological time and no major earth-building movement has taken place since, it follows that the scenery of a large part of Britain is the scenery of glaciation, of a country but recently vacated by ice sheets. It is possible to find parallels in different parts of the world to-day of conditions as they must have existed in Britain at that time and so to study the processes actually at work as well as their after effects. The great ice-sheet which for long covered Scotland must have been broadly comparable with that which now covers Greenland : the tongues of ice which spread from it and from other ice-caps on to lower ground are paralleled by the tongues emanating from Alaskan ice-fields of to-day : valley glaciers of the type which moulded the valleys of north Wales, of the Lake District and of much of Scotland can still be studied at work in the Alps, whilst the great Scandinavian ice-sheet which partly floated across the shallow North Sea to impinge on the coasts of eastern England must have been very like that which floats on the waters of the Antarctic Ross Sea. Southern England beyond the margins of the ice must have had climatic and vegetation conditions comparable with those in the tundra regions of to-day : where the ice-sheets reached the sea there is no doubt that icebergs broke off and floated away with their load of debris to be deposited as the ice melted on the floor of the seas or on neighbouring coastlands just as they do off Newfoundland at the present time. These comparisons are important, because there are so many who find it difficult to conjure up the vision of an ice-bound Britain or to believe that the scenic features they daily see are in fact the results of ice-action. They find it hard to ascribe the smooth contours of Glen Coe to the work of a valley glacier ; to associate the flat floor of the Vale of York with an ice-dammed lake ; to recognise that the very site of York itself was where a dry morainic ridge crossed the marshy plain which remained even after the lake itself had been released and drained away.

Taking first the highlands of older rocks, the example of Greenland shows that only a few of the higher peaks may project above the surface of the ice. Where this does occur the isolated mountains or *nunataks* are extremely rugged in character and with very jagged outlines because they are sculptured mainly by frost action. However, even the highest mountain in Scotland, Ben Nevis, is rounded in outline, suggesting that the ice-sheets overrode the whole of the Highlands ; this is certainly borne out by the character of the Cairngorms. The jagged frost-sculptured upper part of the peaks of the Snowdon Range in north Wales and of some of the Lake District peaks suggests an earlier release from an ice-cover so that frost action has been longer at work. From the centre of the ice-cap, where there is a great thickness and weight of ice, there is a gradual movement of the ice outwards, it being at the same time reinforced at the centre by further falls of snow consolidating into *nevé* or *firn* and then into ice. Under the great ice-sheets the whole country, except any projecting nunataks, is smoothed —all pre-existing soil as well as sub-soil and loose rocks are swept away. What is left when the ice-sheet melts is a smoothed undulating rock surface with the beds of harder rock standing out as low ridges, the beds of softer rock or lines of weakness and crushed rock scooped out to form hollows. It may be that the pre-glacial landscape had much this form and that the ice has merely done a superficial work of removing loose material and generally smoothing the whole but the scratches developed in even the hardest of the rocks illustrate the erosive power of a great weight of ice reinforced by boulders of rock frozen into its foot. But ice has a power which other agents, such as running water, have not, that of actually scooping out hollows where the material is soft. Hence the glaciated rock surface normally has rocky hollows occupied afterwards by lakes. The lakes may be connected by sluggish rivers but water-logged marshy land is abundant. Some of the older rock surfaces in the Outer Hebrides illustrate these features, even better shown in the far west of Ireland in Connemara.

One of the two major effects of glaciation is thus the sweeping bare of large tracts of the country and the time which has elapsed since has generally been inadequate for the formation of a complete soil mantle. There are large tracts of bare rock : soils which have been formed are characterised by immature profiles, whilst the interference with the pre-glacial drainage has resulted in water-logging over huge tracts and the consequent development of extensive bogs. This, indeed, is the reason for the existence of many of the large areas of blanket bog in Scotland.

The other major result of glaciation is the deposition on lower ground or on marginal areas of the material swept from the highlands. The mantle of glacial deposits which results is conveniently referred to under the comprehensive term " drift." Drift deposits are of the most varied character and although two types may be very closely related in geological origin they may be so different in lithological characters as to give rise to land forms, soils and vegetation of the most contrasted types.

When an ice-sheet advances over a plain to a warmer region (e.g. southwards in Britain) or down a valley, there comes a stage when extension by movement outwards or downwards is balanced by the melting of the ice at the ice-front. This line of equilibrium is often relatively stable for a long period of time but as the ice melts the boulders, stones, sand and finer material it is carrying are dropped and a long mound is gradually built up marking the ice-front, for new debris is constantly being brought down by the slowly moving ice. This is the mode of origin of a *terminal moraine*, which in the case of a large ice-sheet may stretch across many miles of country and in the case of a valley glacier right across the valley. It is usually composed of the coarser material since the melt-waters from the glacier carry away the finer and deposit them as an *outwash fan* of glacial sands and gravels over a wide stretch of country.

With a general amelioration of climatic conditions the glaciers retreat. Successive terminal moraines mark periods of temporary stabilisation during the retreat. In other cases the ice-sheets melt on the spot rather than " retreat " and their load of boulders, stones, sand and mud is deposited in an irregular sheet with little or no stratification or sorting of materials. This is the typical boulder clay and whilst the name suggests boulders in a clay matrix there is every possible variation. Passing over the relatively soft rocks of Lowland Britain the ice scooped up vast quantities of material but the bulk of it was not carried very far so that the boulder clay often has close affinities with the underlying solid rocks. Over the sandy Bunter beds the boulder " clay " is thus predominantly sandy and may have very few boulders. After the ice had passed over outcrops of chalk numerous fragments of chalk are found in the boulder clay and it is easy to see that soils derived from the chalky boulder clay

L. D. STAMP

Winter in Fenland—the flooded "washlands" on the right, the peat fen at a lower level on the left (See page 199)

L. D. STAMP

Soil-creep over folded Culm Measure sandstones, Widemouth Bay, Cornwall (See page 101)

L. D. STAMP

A typical Podsol developed on current-bedded Lower Greensand and supporting pinewood. Wrotham Heath, Kent (See page 93)

have been naturally " limed " and are amongst the best agricultural soils in the country, whereas a geologically very similar boulder clay without the chalk fragments may afford soils which are amongst the most intractable to be found. The melting ice must have released huge quantities of water so that irregular patches of sands and gravels usually occur with boulder clays. It is not difficult to visualise the type of country associated with a boulder clay plain—featureless, gently rolling or hummocky with pools of water or swamps in the hollows and rolling mounds often sandy in character; here and there long sinuous mounds which mark moraines and with which may be associated plains of sterile sand representing the outwash fans. Such, for example, is the geology and relief of much of the Cheshire plain and the small remnant lakes or " meres " there form an attractive feature of the landscape.

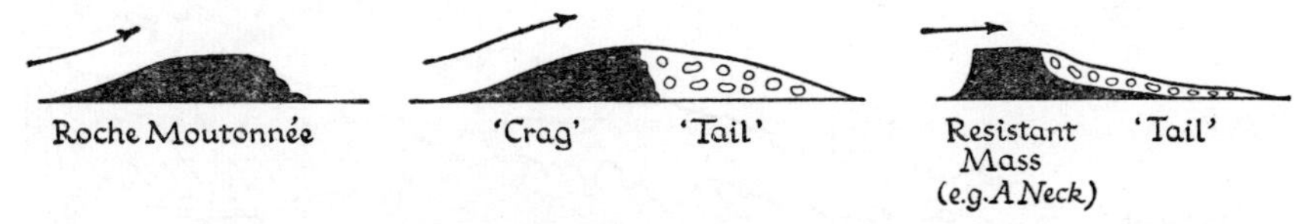

FIG. 34.—Sections showing Crag-and-Tail Structure. The arrows show the direction of movement of the ice-sheet. See page 90 and compare Plate XVB.

Apart from the small meres occupying hollows in the boulder clay plain, larger lakes form an important element in glacial scenery. Those which are due to the actual scooping out of hollows have already been mentioned: there are also those where the water is dammed up by a terminal moraine, especially where the moraine stretches across a valley.

When the ice-sheets advanced across England they played havoc with the pre-existing drainage. In very many cases rivers found their way barred by an ice wall with the result that the waters were ponded back and great lakes were formed. The North Sea Ice for example blocked the outlet of the Humber and a great lake formed behind—a lake to which the appropriate name of Lake Humber has been given—stretching as far south as the present site of Nottingham and as far north as the present site of York. When the waters reached the level of surrounding hills they often established overflow channels which in some cases became permanent water courses but in others were later abandoned and now form " wind-gaps." Where the waters were impounded by a moraine the lakes have remained but in the many cases where they were impounded by ice barriers the lakes themselves disappeared with the melting of the ice but not before the deposition of glacial muds over their floors. Frequently these old glacial lake floors afford extensive plains at the present day and the soils derived from the lake deposits are usually excellent arable soils unless water-logged. Sometimes the margins of the former lakes are clearly marked, even by low cliffs. One of the most spectacular and famous examples of former shore lines of a glacial lake are the " parallel roads of Glen Roy " marking successive levels of an ice-dammed lake.

The greatest extent of an ice sheet is not always marked by a moraine. There is no terminal moraine to mark the greatest extent of

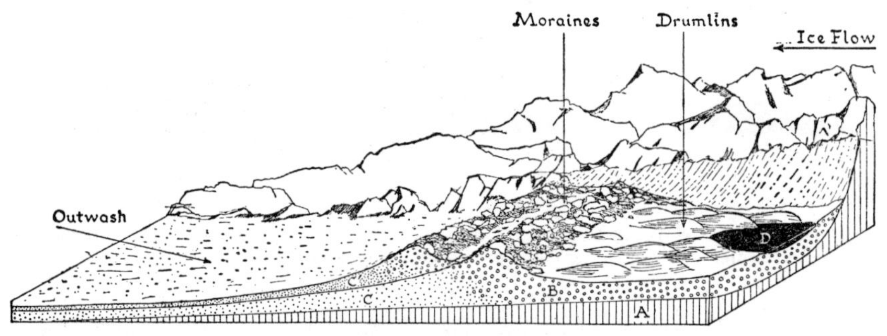

FIG. 35.—Diagram of an Outwash Fan

A represents the solid underlying rocks ; B is boulder clay with a surface of kettle-moraine ; C and C′ are outwash fans from two successive moraines.

the ice sheet in southern England for the ice brought few large boulders right to its tip : tongues of boulder clay reach the latitude of Finchley north of London and then tail out.

Whilst a rolling almost featureless relief characterises the boulder clay lands of East Anglia, in the north of England and southern Scotland a very interesting hummocky topography is the rule. The surface of the boulder clay sheet is in fact a succession of mounds elongated in the direction of the ice flow and consisting also of boulder clay. The origin of these drumlins as they are called, often so surprisingly regular, is not easily explained : it may be that where the ice was heavily charged with debris it became relatively stagnant and the main flow of ice was between these stagnant islands, the material from which was later deposited to form drumlins. It is noteworthy that drumlins tend to occur especially just behind terminal moraines.

From present-day Alpine glaciers it is well known that a stream of water is to be seen emerging from an ice cave at the end of the glacier and that under-ice rivers and lakes are common phenomena. The long winding ridges of sand and gravel (eskers) often seen in Ireland and northern Britain, may have been formed by these under-ice rivers as well as the irregular mounds known as kames (a word which has actually been applied to a variety of surface features) or collectively as kettle moraine. Good examples of country with kettle moraine are seen in the Brampton area of Cumberland or in the Till Valley of Northumberland.

Intermediate between the features which result from the former presence of great ice-sheets on the Highlands and from the spread of those sheets over the lowlands are those associated with the valley glaciers of mountainous regions and which play such a large part in determining the present-day scenery of the Welsh and Cumbrian mountains. The valley glaciers occupied pre-existing river valleys and the main effect of ice action has been to convert the formerly V-shaped valley into one which is U-shaped in section. Further, whereas a river valley develops a regular curve (the curve of the talweg) from its upper to its lower limits, the longitudinal profile in the case of a glaciated valley is irregular because of the power of a glacier to scoop out its bed. A river in its downward course swings from side to side so that a view up a typical river valley reveals a succession of interlocking spurs. A river of ice is less flexible and these spurs are truncated so that the glaciated valley presents a long smooth curve. In lopping off the spurs the valley glacier also cuts off the lower parts of the valleys of tributary streams so that when the ice disappears the water from those side streams drops suddenly from "hanging valleys," often forming waterfalls.

The upper end of a glaciated valley is often, one may say typically, in the form of a cirque (French) or corrie—a basin with steep, often vertical sides, three-quarters of a circle in plan and forming a focus of accumulation of snow which consolidates into ice and then escapes over the lip. It is in fact the birthplace of the glacier and the reservoir from which the glacier is constantly fed. Near the upper rock-wall there is usually a crevasse (the bergschrund) in the ice mass and the ice is constantly attempting to break away from the upper side.

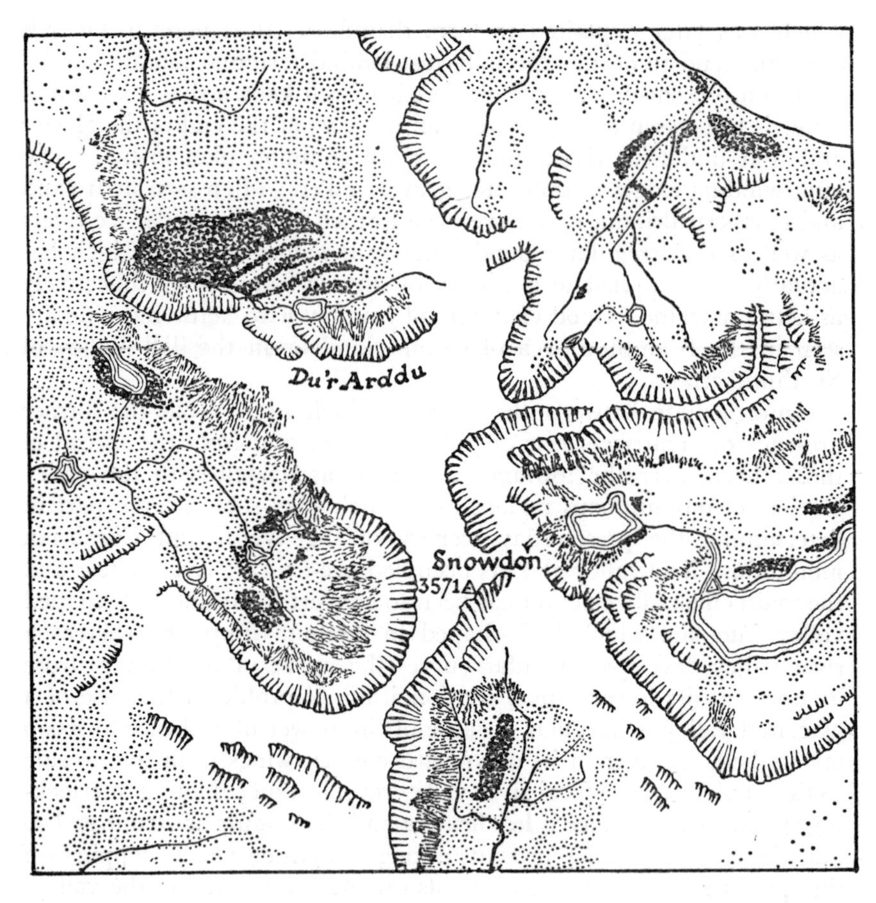

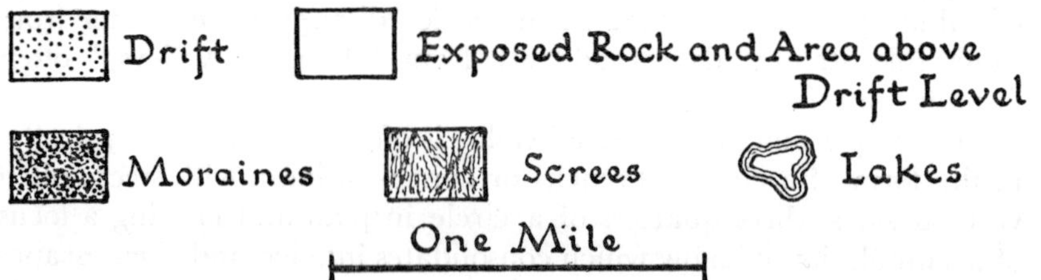

FIG. 36.—Map of the Cirques in the neighbourhood of Snowdon

When it succeeds in so doing, it literally tears away part of the rock face with it and this " plucking " action, especially near the bottom of the bergschrund, accounts for the vertical rock faces of the corrie. When the ice departs the corrie is occupied by an almost circular corrie lake—a very typical feature of the scenery of north Wales—notably of Cader Idris and Snowdon. When cirques from several sides eat into a mountain mass in this way neighbouring ones come to meet in knife-edges or arêtes and in the centre is left a typical pyramidal peak. These features are well illustrated in the Snowdon massif as shown in Plate 8A. See also Fig. 71.

There is a complex series of corries (corrie or coire is the Scottish term; the Welsh cwm has a wider meaning) on the north-east face of Ben Nevis but perhaps the finest group of all in the British Isles is that of the Cuillin Hills of Skye where frost-shattered peaks extend out into knife-edged ridges separating corries with vertical cliff faces up to 1200 feet high. W. V. Lewis, from a study of Icelandic and British examples, has come to the conclusion that summer rain and melt-water, flowing down the face of the rock—between the rock face and the ice—plays a large part in the general process of sapping and plucking whereby blocks are removed from the rock wall and that this goes on independently of the existence of a bergschrund.

Both valley glaciers and ice-sheets when they melt may leave stranded in curious and anomalous positions blocks of rock, often far-travelled and of large size. These erratic or perched blocks are a sure indication of glacial action since no other agent exists which could have transported such masses. Blocks of the distinctive granite of Shap Fell are found over many parts of the northern Pennines and enable the former course of the glaciers to be traced.

An interesting phenomenon seen on hill slopes near ice-sheets on glaciers is the formation, by the process noted above as solifluction (p. 20), of a semi-frozen sludge of hardened snow, ice and stones which moves slowly downhill and when the ice melts leaves a very irregular jumble of angular rocks filling valleys or spread over low ground. The deposit known as "Head"—a rubble drift of local rocks mantling the slopes and filling the lower parts of many valleys in the south-west of England (which was never directly covered by ice-sheets)—is of this origin and so is the " Coombe Rock," a chalky rubble which fills many of the valleys in the North and South Downs. It is not easy to explain some of the steep-sided valleys in the chalklands of southern Britain such as the Devil's Dyke near Brighton and a possible explanation is that they are comparable to corries with local sludge glaciers rather than true ice glaciers.

Some interesting features of glacial erosion in this country are associated especially with the central lowland of Scotland where ice streams were escaping outwards towards both the west and to the North Sea. They found in the way hard plugs of volcanic rock such as that forming the Castle Rock on which Edinburgh Castle now stands. The ice swept away the loose material in its path but in the lea of the rock mass even relatively soft material was protected so that the crag has a " tail " partly of rock, partly of superficial deposits. This crag-and-tail structure is perfectly illustrated by the Edinburgh example where the gentle slope up to the Castle from the eastern side represents the " tail." Smaller rounded smoothed rock masses are known as roches moutonnés because of their resemblance from a distance to sheep at rest. These are, indeed, very common in glaciated areas and are now seen in many parts of North Wales.

SOILS

FOR MORE than a century after the emergence of geology as a serious scientific study there still remained a no-man's-land—the study of the soil. The agriculturalist was, indeed always has been, of necessity vitally interested in the soil as the medium in which his plants grew, and he accordingly regarded the study of soils as a means to the very important end of improving his cultivation. To the geologist the soil is even a worse nuisance than the superficial drift deposits: it hides the rocks below and its presence may involve the necessity even of digging pits to see what is underneath, and when mapping the soil must be imagined completely removed. In this country the nature of the soil is very closely related, generally speaking, to the nature of the underlying rocks and it became usual to talk about "Chalk soils," "London Clay soils," "Lower Greensand soils" and so on, thus emphasising this close association. This tended, indeed, to over-emphasise the connection and to hinder the study of the soil for its own sake. To find soils mapped as such one has to go back to the days before the development of scientific geology. When the "Board of Agriculture and Internal Improvement" (the unofficial predecessor of the Ministry of Agriculture) was set up in 1793 one of its first acts was to call for reports on the state of agriculture in each of the counties. In due course these were revised and published as a series of octavo volumes under the title "A General View of the Agriculture of . . ." and most of them have a hand coloured soil map of the county. These are still the only soil maps which exist of many of the counties of Britain.

The emergence of soil science or "pedology" as a separate branch of science is mainly within the past few decades, at least so far as this country is concerned. The Soil Surveys of England and Wales and of Scotland were established shortly before the outbreak of the Second World War. The first one-inch sheets and memoirs were published in 1954.

In a few areas the Geological Survey added an investigation of soils to its regular mapping work. Two coloured maps on the one-inch scale covering a large part of Ayrshire were published and in the Brampton

(Cumberland) sheet memoir is a black-and-white map of part of the Eden Valley. The basis of the soil classification used for these maps is a textural one : the experimental nature of the work is clear from the fact that three different classifications are used on these three maps, but broadly the purpose was to separate sands, light, medium and heavy loams, clays and peats. The work of agriculturalists and chemists is well illustrated by the now classic work by the late Sir Daniel Hall and Sir John Russell on the *Agriculture and Soils of Kent, Surrey and Sussex.* They combined detailed chemical with mechanical analyses but their work was related to the practice of agriculture and to obtain their samples of soil the surface layers were well mixed—just as a plough would do. Though published as recently as 1912, the procedure would not now be regarded as the correct one.

Pedology in the modern sense may be described as the philosophical approach to the subject—the soil is studied as soil. It was the Russians, headed by Glinka, who developed the concept that it is climate rather than parent rock material which determines the main characteristics of soils and this is undoubtedly true in so far as the great world groups are concerned. British soils all belong to but two or three of these world groups since the British Isles lie within one major climatic region—one in which rainfall exceeds evaporation so that the dominant movement of water in the soil is from the surface *downwards* and there is a natural tendency for soluble salts to be washed out of the surface layers and carried down to be redeposited at lower levels. This process of *leaching* is the opposite of that occurring in dry climates where water is drawn upwards and evaporates leaving a deposit of salts in the surface layers—hence the saline soils of desert lands.

It is not difficult to see why the basis of modern soil study is the *soil profile*, as a section through the soil is called, and why some pedologists prefer to consider only " natural " soils since the efforts of man in ploughing and so turning over the surface layers is actually to *prevent* the action of normal soil-forming processes. If these processes are allowed to go on undisturbed, the climate of Britain results normally in the differentiation of three distinct horizons in the soil and it is these which make up the soil profile. The upper or A horizon is that subjected to downward washing or leaching ; the finer materials are washed down mechanically to the lower or B-horizon and soluble salts are dissolved, also to be washed downwards and perhaps there re-deposited so that the B-horizon is one of secondary enrichment. The

L. D. STAMP

A black, peaty meadow soil, overlying chalk and caused by bad drainage along Mill Brook near South Moreton, Berkshire. (See pages 94-95)

PLATE 12

GEOLOGICAL MUSEUM

Colour photograph of a model of the southern part of the Cumberland Coalfield, looking southwards with St. Bees Head in the distance (See page 120 and page 244)

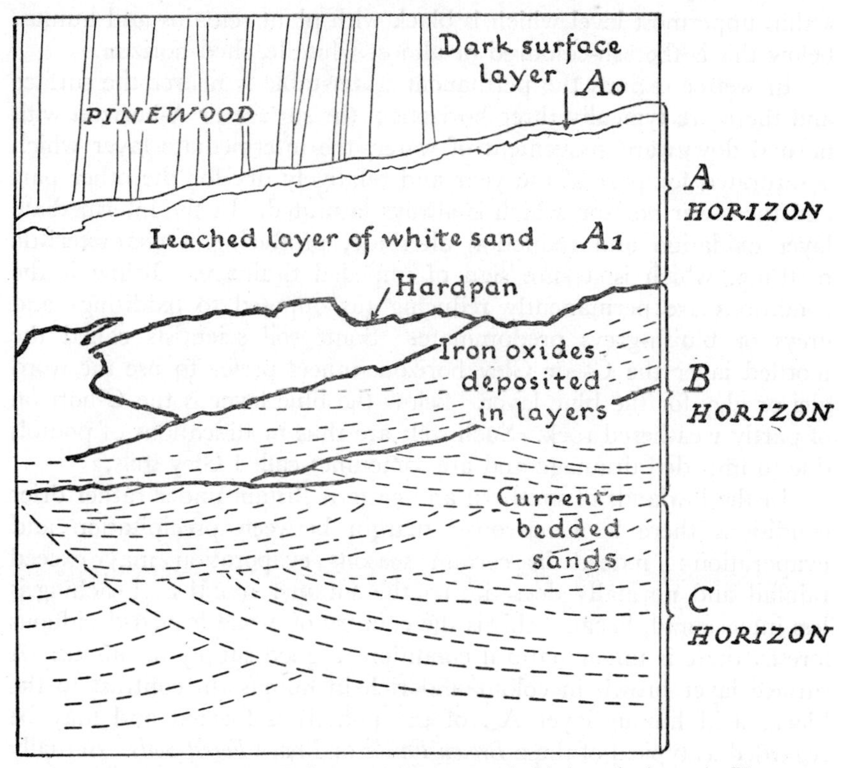

FIG. 37.—Diagram of the Podsol overlying current-bedded sands shown in Plate 10 Many soil scientists restrict A_0 to the layer of unaltered surface litter, calling the dark layer A_1 and the leached A_2

B-horizon is often stained brown by oxides of iron and these may be deposited in sufficient quantity to produce a " hard-pan "—a deleterious feature of many soils which prevents the penetration of plant roots and needs to be broken up before trees can be planted or before cultivation of crops can be successful. Below the B-horizon is the C-horizon which is the parent rock material, broken up it may be by weathering and root action. The soil profile just described is that of a typical *podsol.* It may be regarded as normal in Britain except in the wettest parts and in the drier parts of the south and east or where the sub-soil materials are of a special character (*e.g.* limestone). Podsols may be seen in their most characteristic form in sandy land such as the Bagshot heaths of Surrey. There the growth of surface vegetation gives rise to a thin uppermost level which is black with plant remains and humus, below this is the ash-coloured or almost white leached horizon.

In wetter regions the permanent water-table is nearer the surface and there are typically three horizons ; the surface or A-horizon with normal downward movement of water, the intermediate layer which is saturated for part of the year and relatively dry for the other part and the lower horizon which is always saturated. In the intermediate layer oxidation and reduction alternate, producing a characteristic mottling, which is a sure sign of impeded drainage. Below it the conditions are permanently reducing (as opposed to oxidising) and greys or bluish-greys predominate. Some soil scientists called the mottled layer the G- or Gley horizon, others prefer to use the word glei or gley for the blue layer. Below the blue layer is the C-horizon of partly weathered rock. Such soils are thus modifications of podsols due to impeded drainage and are sometimes called Gley soils.

In the lowlands of southern and eastern Britain under rather drier conditions there is a narrower margin between precipitation and evaporation : indeed in certain seasons evaporation may exceed rainfall and normally does during the summer months. Leaching is less intense and, because this is the country of broad-leaved deciduous forests, there is under natural conditions a good supply of humus. A surface layer brown in colour and rich in humus (in contrast to the black, acid humus layer A_0. of the podsol) is formed and may be regarded as typical of these *Brown Earths* or *Brown Forest Soils*. Actually however there are so few parts where these soils occur that have not been ploughed or drained or artificially " managed " in some way or other that it is difficult to say how far these are " natural " soils.

An interesting group of soils is developed on limestones—notably on the chalk. Sometimes, curiously enough, the surface layer is so leached of lime that it is markedly acid in reaction. More commonly lime is in excess in soils developed on limestone and such soils are known as " rendzinas."

Over a large part of Highland Britain, especially in areas from which all loose rock as well as previously existing soil was removed during the Ice Age, there has been insufficient time for the weathering of the rocks and the formation of a complete profile. These are termed " immature soils." In other cases soil erosion has washed away the upper levels and one may find soils with truncated profiles.

From this brief account of the character and development of the soil profile it will be realised how fundamentally opposed to the modern approach to the study was the old method of *mixing* the layers of the soil and subjecting the mixture to a mechanical and chemical analysis. The modern classification within the great groups is based on the description of the soil profile. The main unit of classification is the soil series defined as " a group of soils developed from the same or similar parent materials under similar conditions and showing similar profile conditions." Soil series are usually named after places where they were first studied or are typically developed. Within each soil series there may be a considerable range of texture which is important ecologically, and there may be shallow and deep " phases " which are equally significant.

Despite these changes in the standpoint from which soils are studied, texture remains of fundamental importance—the vegetation of a clay soil is essentially different from that of a sandy soil. Texture depends primarily on the size of the mineral particles and there is now international agreement on the nomenclature of particle-sizes thus :

Diameter of particles	above 2 mm.	—stones
	0.2 to 2 mm.	—coarse sand
	0.02 to 0.2 mm.	—fine sand
	0.002 to 0.02 mm.	—silt
	under 0.002 mm.	—clay

Sandy soils usually contain more than 60 per cent of coarse and fine sand and less than 10 per cent of clay. Owing to the large size of the particles they are readily permeable to both air and water. Roots not only penetrate such soils easily but they can also " breathe " easily. On the other hand unless there is an underlying clay layer water may drain away so rapidly that shallow rooted plants may be left dry. As the farmer would say, pastures on such soils are liable to " burn." Moreover, plant foods are liable to be leached away and when cultivated such soils need constant manuring—they are " hungry " soils.

Loamy soils or *loams* contain a smaller percentage of sand and a higher one of silt and clay. They are easily penetrated by roots but afford a firm hold ; they are retentive of moisture and plant food and are not only the best agricultural soils but where not cultivated support a varied flora. Naturally they are, in fact, almost entirely cultivated.

Clayey soils or *clays* contain a high proportion of silt and clay ; when dry they bake into hard clods, when wet they form a sticky mass, when very wet they form a sludge or viscous mass. Only a limited number of plants are able to flourish under these difficult conditions, with water-logging and lack of aeration to the roots for part of the year and the damage to roots by their being torn apart when the soil bakes and cracks in the summer.

It is, however, a fatal error to rely too exclusively on the mechanical analysis of any soil without considering the ways in which the constituent particles are arranged. It is this which constitutes the "structure" of the soil. Under certain circumstances the particles collect together to form "crumbs" and hence to leave air and water space between the crumbs. If a sample of well-cultivated soil is taken from a garden and pressed tightly into a jar the crumb structure is not destroyed. But if it is then heated and dried and the crumb structure destroyed it will pack away into a much smaller space, even after allowing for the loss of water. It is mainly the maintenance of the soil-structure, especially this crumb structure, which the farmer means when he talks about "keeping the land in good heart" and "maintaining a fine tilth." The structure of the soil depends to a large extent on the amorphous organic material conveniently called *humus.* In many soils it is possible to pick out fragments of roots, stems, leaves, etc. but the larger proportion of the remains of plants has been converted by the action of soil-bacteria and other micro-organisms into the amorphous, usually but not necessarily dark-coloured, humus, a substance apparently colloidal in nature and extremely difficult to analyse chemically. It is often in such close association with the clay fraction of the soil that it is usual to talk about the clay-humus association or the colloidal complex since clay itself is so fine-grained as to act almost as a colloid. It is clearly of the utmost importance to maintain the humus content of the soil—and it is here that organic manures perform a dual function. They must not be regarded as merely restoring to the soil certain chemical compounds needed by plants (a function as well if not better performed by chemical fertilisers or "artificials"), since they maintain the all-important structure of the soil which chemical manures, speaking generally, do not.

When organic matter—remains of plants and animals as in ordinary farmyard manure—is added to soil it may suffer one of two fates. It may be oxidised, a process equivalent to burning in that, with the addition of oxygen, most of the organic matter is converted into gases.

THE COAST AT FOLKESTONE, KENT, LOOKING EASTWARDS AERO-PICTORIAL

The coastwise drift of shingle is obviously to the east as shown by the groynes. The breakwater on the west of the harbour has acted as a gigantic groyne and the land at the foot of the cliff has resulted since it was built. In the distance is the landslip area of the Warren, where masses of chalk slip seawards over Gault clay. In the far distance are Shakespeare Cliff and Dover Harbour

Plate X

BLAKENEY POINT, NORFOLK, LOOKING EAST NORFOLK AND NORWICH AERO CLUB

To illustrate the building-up of a shingle and sand spit by a drift of material, in this case from the east

These gases, mainly carbon dioxide and ammonia, may be dissipated into the air though the ammonia, by further oxidation within the soil, may be converted into nitrates and become directly available as plant food. In the same way by oxidation other constituents of the organic matter may become readily reavailable as plant food to growing plants. Up to a certain point oxidation is thus good and it is one of the results of ploughing to bring air into contact with the organic matter and promote oxidation. But if oxidation is too rapid or allowed to proceed too far the soil may be rendered completely sterile. Repeated ploughings have been proved, especially in tropical countries, to promote oxidation to such an extent that a good soil has been completely ruined in a short space of time. On the other hand the organic matter in the soil may be converted into humus by the process of humification. Some of the organisms which carry out this change work entirely out of contact with air (anaerobic organisms), others (aerobic) work under more normal conditions in the presence of air. In a healthy soil it is essential to maintain a balance between oxidation and humification. A soil which is very rich in organic matter but which is properly aerated by good drainage provides such a balance ; hence the very valuable agricultural qualities of drained mosses and fens—the so-called mild peats. On the other hand with excess of moisture, as in regions of the British Isles with very heavy rainfall combined with bad drainage and consequent poor aeration of the soils, there accumulates an excess of humic acids and hence " acid " peats.

The mineral particles of a soil consist for the most part of quartz grains or of other minerals which are " stable," i.e. insoluble in water or the ordinary soil acids and so are actually or relatively inert. Limestone (calcium carbonate, Ca CO_3) is a rock of widespread occurrence in nature and often forms the bulk of the sub-soil and the parent rock of many soils. Limestone is different from other parent materials in that it is dissolved in acids. It is slowly soluble even by weak soil acids. It combines with the humic acids and so neutralises them. In all the wetter parts of this country and in very many soils the natural tendency is for the accumulation of an excess of soil acids—hence the almost universal importance of liming. It is possible to find soils overlying limestone from which all lime has been leached out and hence the apparent anomaly of a lime-hating (calcifuge or calciphobe) flora sometimes found in limestone country. Where this is the case and the soil is shallow, deep ploughing may be a sufficient remedy by turning up fragments of limestone from the subsoil. In the old days an immense amount of " chalking " was carried out, especially in southern England—the spreading of chalk dug from shallow pits over clay lands. This had the advantage that the lumps of chalk were attacked slowly by soil acids and the effect of a good chalking might last half a century or more, whereas the spreading of ground limestone (favoured in modern days because it can be spread quickly and evenly by machinery) and still more of lime (quick or slaked) whilst more immediate in its effect is far less lasting. Although in modern geological literature the term " marl " is used generally for a rock containing some lime, the old practice of " marling " meant the spreading of clay or " marl " on light sandy land in order to give the soil more body—in reality to attempt to convert a sandy soil into the much more valuable loam.

Lime in a soil not only neutralises the excess acids but it forms what are known as exchangeable lime compounds—compounds of lime with the organic acids, resulting in the release of water, which can be drained away, and of carbon dioxide. Apart from this, lime has a most important effect on soil structure by promoting " flocculation " whereby fine particles of clay are collected together in groups and a crumb structure developed in the soil thus rendering clay soils more friable or workable.

Although the chemical composition of a soil, especially the salts dissolved in soil water and so available as plant food, is important, any mineral deficiencies are relatively easily remedied by balanced manuring with artificials. This is an easier process than altering the texture or structure of a soil. Two universal needs of plants are compounds containing combined nitrogen—loosely called by agriculturalists nitrates or nitrogenous manures—and compounds containing phosphorus (which is a constituent of protoplasm and hence of all living matter)—loosely called phosphates by the agriculturalists. The required nitrogen can be supplied by such extremely soluble mineral salts as Chilean nitrates or ammonium sulphate but the supply in the soil can be built up by the cultivation of leguminous plants such as clovers, in the root nodules of which live those bacteria capable of " fixing " atmospheric nitrogen. The necessary phosphorus is supplied in the forms of bone-meal, basic slag and superphosphates. But both nitrates and phosphates are contained in good farmyard manure and the presence of micro-organisms in the latter is an additional advantage. It is one of our crimes as a nation that we tip immense quantities of valuable manure into the sea in the form of sewage. On light soils there is much to be said for feeding concentrates to animals and allowing their nitrogen- and phosphorus-rich droppings to fertilise the soil at the same time as they pack together the particles of soil by the treading action of the " golden hoof." Heavier soils, such as clays, on the other hand, are rendered less tractable by this treatment.

Although plants require also sulphur, magnesium and iron, in practice manuring is concerned mainly if not exclusively with making up deficiencies in nitrogen and phosphorus and, of course, lime. In recent years there has been growing appreciation of the important part played by elements which may be present in such small quantities that they remain undetected by ordinary methods of chemical analysis. These so-called " trace " elements include manganese, molybdenum, boron, iodine, copper and zinc. Deficiencies in one or other are responsible for many plant diseases. The quantities involved are often extremely minute ; molybdenum is essential to the growth of the tomato plant but it has been found that one part in 100,000,000 in the nutrient solution is adequate. On the other hand it is now known that an excess of molybdenum in the " teart " pastures of Somerset caused an acute form of diarrhoea amongst the cattle pastured on them—an illness only made worse by improving the quality of the pastures because improvement meant more clover which absorbed more molybdenum.

It will be clear that the soils of the greater part of the British Isles, certainly of Lowland Britain, have been profoundly modified by the hand of man and that even in areas apparently entirely uncultivated a good chalking or a good marling carried out a century or more ago may be a dominant influence in determining the present character of the vegetation.

It has been emphasised that the actual formation of soil from the parent material is mainly the work of circulating (or it may be non-circulating) waters and it follows that drainage operations, local or regional, surface or sub-surface, have a profound effect on soil-forming processes. The natural fertility as well as the productivity of a soil are even more dependent on water conditions than on natural content of plant-foods.

It is common knowledge that the level of water in a well fluctuates with the seasons but that there is a level below which the water normally never falls. This is the level of the " permanent water-table " in the surrounding rocks. The height of the permanent water-table varies considerably; in limestone uplands it may be far below the surface: in riverside meadows and marshes it may only be a few inches below. It may actually be *above* the land surface, in which case a pond or swamp conditions result—as in the reed-swamps on the borders of the Norfolk Broads. Where the water-table rises above the land surface only in winter we find swamp conditions only then, though in summer the water-table may be but a few inches below the surface. It is in soils where the water-table is nearly at the surface at one season and a foot or two lower at the other that there occurs the development of the gley horizon already described.

It is important to note what happens in a soil above the level of the permanent water-table. Even when the soil is "dry," a film of water remains round the grains which is only given up when the soil is heated in the laboratory to temperatures above those to which it is subjected in nature. When such a dry soil, using dry in the field sense, is moistened by a shower of rain it will soak up water until it is saturated to its "field capacity," which varies greatly according to the texture and consequent porosity of the soil. Unless the soil is a very fine-grained clay soil there will still remain plenty of pore-space and the soil is well aerated. Any further water will simply drain through and the soil, like its rock counterparts, it is said to be permeable. An ideal soil, indeed a good soil, must be porous and able to hold a good quantity of water; it must also be permeable so that excess water drains through and does not cause water-logging. It is clear that porosity and permeability are two very different properties.

Soil drainage is of more than one kind and has more than one purpose. Surface drainage is designed to carry off excess of surface water; sub-soil drainage, so widely effected by pipes, mole-ploughing or the digging of deep trenches into which the water seeps laterally, is designed to lower the water-table and to promote sub-surface drainage. Then the relationship between vegetation cover and soil drainage is far from being sufficiently realised. Where land covered with forest is in a heavy rainfall region a considerable quantity of the moisture which falls never reaches the ground but is evaporated from the leaves and branches of the trees. Further, a large proportion of that which soaks into the ground is absorbed by the trees and lost by normal transpiration. When the forest is felled *all* the rain reaches the ground and the low-growing plants such as *Sphagnum* lose far less per acre by

A PRE-GLACIAL ROCK PLATFORM, NORTH COAST OF ISLAY H.M. GEOLOGICAL SURVEY

This view illustrates the action of the sea in cutting almost level platforms, which may afterwards be elevated. The line of old sea cliffs on the left is very clearly marked

Plate XII

THE CULBIN SAND DUNES, MORAYSHIRE H.M. GEOLOGICAL SURVEY

Moving sand, being blown from the right, is overwhelming sand already partly fixed by marram grass

evaporation than forest trees. It has been well said that for much of the wettest west of Britain the alternatives are forest or bog and man is the deciding factor.

In recent years attention has perforce been drawn to the world-wide dangers of soil-erosion. In so many directions experience in the British Isles is no sure guide in other lands. Just as there is no surer way of ruining a tropical soil than by promoting that oxidation which results from frequent ploughing—a treatment which would be beneficial to many British soils—so there is no surer way of losing a soil in a land of heavy downpours than to practise deep ploughing with long straight furrows running up and down hill which one can use with relatively little danger, even with advantage, in England. The removal of the natural cover grass exposes the American prairie soils to all the terrors of wind erosion and for the same reason clean weeding, which is a virtue on an English farm, may be a most dangerous vice on a tropical one in West Africa. However, because of our relative immunity it would be wrong to imagine that soil erosion is not found in Britain. The turbulence of any stream after heavy rain should make such a supposition demonstrably untrue and it is easy to see the downward wash of fine particles of soil in almost any sloping field after heavy rain. Truncated soil profiles on higher lands tell the same story and " hill-creep " or " soil creep " can similarly be seen in most hillside sections (see Plate 9B). The much greater thickness of soil in most valleys, for example in the dry valleys of chalk lands, is further evidence. Though farmers know that certain types of soil will " blow," it was especially in the years of the war-time ploughing campaign that wind erosion over large areas of such valuable lands as Fenland became serious. Apparently unwanted or unnecessary belts of trees removed to make larger fields suitable for mechanised cultivation were hastily realigned and attempts made to restore them. The power of wind in promoting soil erosion once the protective vegetation cover is removed is well seen in the " blow-outs " of Breckland, whilst the movement of unconsolidated sand-dunes and their ability to overwhelm forests is only too well known from the Culbin sands of the Morayshire coast (see Plate XII) and elsewhere round the coast of Britain. An unusual case of gully erosion is shown in Plate XVIII.

In the preceding pages we have attempted to indicate the wide variety of soils present in Britain and to give some account of their origins and their characteristics. It is clear that they vary greatly in

reasons that they have such an important effect on scenery. The better soils are cultivated ; it is the poorer soils which have remained largely untilled and so covered with woodland, moorland or rough grazing. Older writers express hatred for the " rascally heaths " and " wastes " on poor soil and Cobbett's strictures on Hindhead are well known. Yet to the walker and to the cyclist escaping from London these have become the havens of wild nature which, above all other type of country, he now seeks. Similarly, it is the very poverty of the soil on the steeper chalk slopes or on Carboniferous Limestone which has permitted the survival of typical lime-loving plants.

GEOGRAPHICAL EVOLUTION

AS WE HAVE already observed (p. 10), in the early days of the development of geological science a fierce controversy raged between two schools of thought. There were those who averred that the sequence of the rocks with their entombed fossils could only be explained by a series of catastrophes—of which Noah's flood was presumed to be one—vastly different from the ordinary processes which are observed on the surface of the earth at the present day. On the other hand there were those who saw the interpretation of the record of the rocks in the processes of erosion and deposition which can be observed taking place now and who found it unnecessary to postulate violent convulsions and phenomena different from those still to be observed. Gradually the latter school came to hold the field and the modern concept of geology was finally firmly established with the publication of Charles Lyell's *Principles of Geology* in three volumes issued between 1830 and 1833 and followed by the even more famous *Elements of Geology* in 1838. Lyell was the son of a Forfarshire land owner and botanist and was born in 1797. He was destined for a legal career but came at Oxford under the influence of Dr. William Buckland to such good purpose that he read his first paper to the Geological Society in 1822, became one of its Honorary Secretaries in 1823 and was elected a Fellow of the Royal Society in 1826 at the age of twenty-eight. Lyell's two great works passed through many editions and it was in his third famous volume on *The Antiquity of Man*, published in 1863, that he first indicated his adhesion to Darwin's doctrines in the *Origin of Species* (1859). Despite the vast volume of geological research over the past hundred years the bulk of what Lyell wrote remains true ; indeed there has recently been a swing back to Lyellian teachings which had been partially abandoned. We recognise that to-day we are fortunately living in a comparatively quiet period of the earth's history when major earthquakes which are the indication of mountain-building movements are rare and when volcanic activity, though widespread, is rarely violent. But otherwise past conditions on the earth's surface were much as they are to-day : there were some areas of high relief where all the many processes of sub-aerial denudation were active, other areas worn almost down to base level ; there were areas of heavy rainfall, others of desert aridity just as there are to-day. It is true that some questions relative to the past remain unanswered or at the best partly answered. One such is relative to past climates—there seem to have been several periods when the greater part of the world was in the grip of an ice age and others when remains of vegetation suggest that the whole surface enjoyed relatively warm conditions. But even so it is possible to match the conditions as they existed in the Ice Ages with those found in the colder parts of the world of to-day. In explanation of some phenomena it has been urged that at times the compositon of the atmosphere may have been different from what it is at the present day and that a larger proportion of carbon dioxide may have encouraged a more luxuriant growth of vegetation in Coal Measure times. It has also been urged that the seas must have been less salt but, by and large, what we can reconstruct from the evidence of the rocks and their fossils does not suggest wide divergencies from conditions as they are found in one part or another of the surface at the present day.

All this is really equivalent to saying that geology—historical geology or the study of stratigraphy—is geographical evolution and that we can best understand the past if we attempt to reconstruct the geography of the past and to match the conditions revealed by the rocks with conditions to be found in different parts of the world to-day. Thus it is possible to visualise clearly the salt lakes and desert flats of Permian and Triassic times by seeing how closely the redness of the rocks, the wind-polished sand grains, the shattering of crags by the disruptive force of the sun's rays and other features can be paralleled in the deserts of the Sahara, Persia or Chile at the present time.

Naturally an immense amount of patient field work and laboratory investigation is needed before one can piece together anything approaching a full picture of the geography of a distant geological period. The earlier in the earth's history the more fragmentary the picture is likely to be, and the more doubt must still remain of the validity of deductions. Nevertheless the results are fascinating for, traced from one period to the next, one begins to see how the changing distribution of land and sea, the relative positions of mountains and plains and the various stages in the cycles of weathering have all left their records in the rocks.

When I first began lecturing in geology in the University of London at King's College in the autumn of 1919 it was my ambition to arouse the interest of students by seeking to reconstruct for them the geography of the British Isles as it must have been at each of the great geological periods and this is the approach which I adopted in my book, *An Introduction to Stratigraphy*, which I wrote during the leisure of my first voyage to Burma in 1921. The geographical reconstruction of the earlier Palaeozoic periods is naturally sketchy in the extreme and I well remember that I tried out my method by writing first the chapter on the Permian, by when the picture has become clearer. Later writers, using this method, have experienced the same difficulty. Professor L. J. Wills, in his *Physiographical Evolution of Britain*, starts with the Permian and Trias and Dr. A. K. Wells in his *Outline of Historical Geology* (1938) starts with the Cambrian but returns later to deal with the pre-Cambrian.

As it is the main purpose of this present book to trace the development of the physical features and environments of this country as they are to-day, an attempt will be made to trace the gradual evolution of its geography through the ages. It will be appreciated that the great mountain-building movements of the past have resulted in the intense folding of the rocks in many areas and it is thus clear that rocks which now occupy but a small space—such as the intensely folded Devonian and Carboniferous deposits of Devon and Cornwall—were laid down over a much greater actual area. But in attempting to draw the outlines of land and sea as they existed in past ages it is only possible to relate them to the *present-day* coastline of the British Isles. It is necessary always to remember this in interpreting the maps which follow. What is shown on these maps as a narrow sea was often one of great width but of which the deposits have been compressed by folding.

Starting with the Cambrian—for it is virtually impossible even to hazard a guess as to the distribution of land and sea in any of the pre-Cambrian periods in Britain—a striking fact is the contrast between the Cambrian rocks and fossils of the extreme north of Scotland (notably in the Durness Limestone) and those of Wales. At the same time there are marked affinities between the Scottish faunas and those of the northern province of Canada and the United States and also between the Welsh faunas and those of the eastern or Appalachian province of the United States. Largely on this evidence there has been postulated the existence of a deep ocean trough stretching across what is now the

Atlantic Ocean roughly from west-south-west to east-north-east. Migration was possible along the *shores* of this ocean much more easily than *across* it. At the same time, though some of the Cambrian deposits in Wales are fine-grained shales, which may have been deposited in

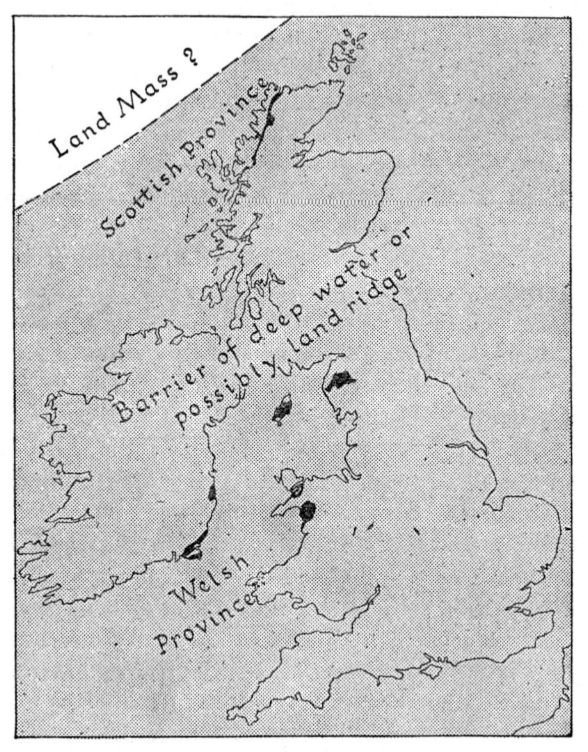

Fig. 38.—The Geography of Cambrian times

In this and all similar palaeogeographical maps which follow, the present-day outcrops of rocks of each age concerned are shown in black. The former extent of sea is shown shaded, former land areas are shown white.

deep water, there are sandstones with faunas which suggest shallower waters. Thus the margin of the ocean may not have been far north of the British Isles and there was probably a land barrier separating the Americo-British ocean from that in which the Cambrian Beds of Bohemia, with very different fossils, were deposited.

The Cambrian rocks of north-west Scotland are involved in the very intense folding of the Highlands which is believed to be mainly

Caledonian in age so that there is no inherent impossibility in the concept of a great oceanic trough occupying what is now central and southern Scotland. There is, it is true, no positive evidence and it may be that the differences between the Cambrian faunas of Scotland and

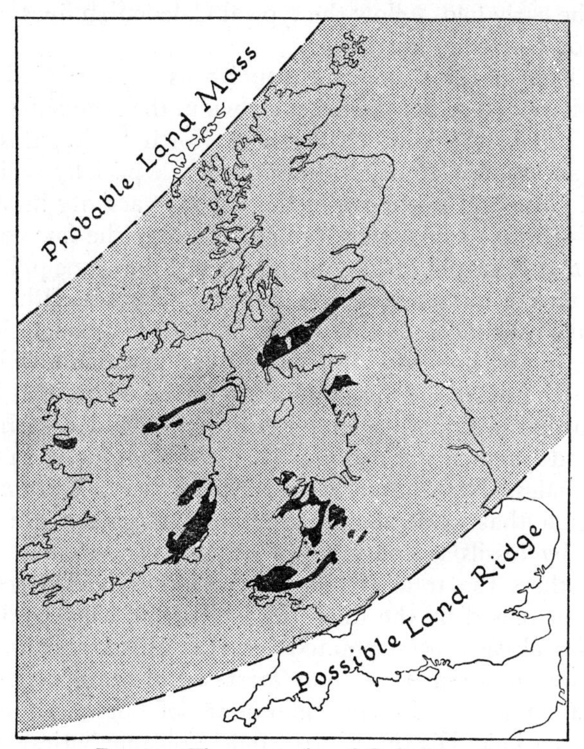

FIG. 39.—The geography of Ordovician times

those of Wales are to be explained by postulating a land barrier. Against this are the characteristics of the Skiddaw Slates of the Lake District and the Manx Slates of the Isle of Man which are believed to be Cambrian in the main and are fine-grained rocks suggesting relatively deep water. Further, supposedly Cambrian rocks—rather of Welsh type—occur in Kincardineshire, and there is, in fact, no evidence for a land barrier.

If we accept the basal idea of a major oceanic trough across the British Isles in Cambrian times, the geological history of the Ordovician and Silurian fits into the picture—it is essentially the history of the "buckling" of the trough, with consequent shallowing in different parts—even partial emergence above sea-level so that there are local unconformities and areas where the rock sequence is incomplete. The buckling was in fact the prelude to the great Caledonian folding and in general the early folds follow the typical Caledonian lines, with their axes from south-west to north-east.

The buckling of the Cambrian trough was accompanied by much volcanic activity. As A. K. Wells has shown in the Rhobell Fawr mass, on the margins of the Harlech dome in North Wales, this volcanic activity began in late Cambrian times but it is especially typical of the Ordovician. Much, if not most, of the volcanic activity in Wales was sub-marine—sheets of lava were poured out over the floor of the sea, sometimes in curious pillow-like masses (hence the name pillow-lavas), but very often the molten rock was forced as thin sheets between the recently formed sediments so that it is often hard to distinguish between true lava flows and intrusive sills. Where the volcanic activity was of an explosive character the dust and ashes blown out settled back into the water so that layers of volcanic ash are interbedded with ordinary sediments and fragments of ash and lava are often found in the sandstones and shales. Naturally there were local foci of volcanic activity with the result that while the sequence of Ordovician time-zones as determined by fossils is constant, in some areas volcanic rocks are absent, in others the total thickness of the Ordovician sequence is enormously increased by the thick masses of lava, ashes and intrusive rocks. It has already been pointed out that these igneous rocks are more resistant to weathering than most sedimentary rocks and are responsible for many of the chief mountains of England and Wales—the heart of the Lake District (Borrowdale Volcanic Series), Snowdon, Cader Idris, the Arrans and so on.

The Ordovician sequence has been divided into some twelve or thirteen zones defined by their graptolite fossils, which give precision to the older three-fold division which was (in descending sequence) :

Bala Group (=Caradoc group of Shropshire or Hartfell of southern Scotland).

Llandeilo Group (=Glenkiln of southern Scotland).

Arenig Group (= Skiddaw Group or upper part of Skiddaw Slate).

The separation of some higher Arenig Beds and their grouping with some lower Llandeilo to give a Llanvirn Group results in the now commonly used division into :

Plate XIII

PIPES IN CHALK, NEAR CANTERBURY, KENT

L. D. STAMP

Solution hollows in the soft white limestone have become filled with slipped mass of red-brown clay-with-flints

Plate XIV

GEOGRAPHICAL PUBNS.

THE DRIPPING WELL, NEAR KNARESBOROUGH, YORKS

The water which flows over this cliff of Magnesian Limestone is so highly charged with calcium bicarbonate that it deposits a thin film of crystalline limestone over any object on which it drips and from which the water evaporates

Arenigian, Llanvirnian, Llandeilian, Caradocian and Ashgillian; the two latter being the lower and upper parts respectively of the old Bala Group.

The folding to which reference has previously been made was already in evidence at the end of Cambrian times so that there is in some areas an abrupt base to the Ordovician sediments in the coarse Arenig Grit—undoubtedly formed where the water was relatively shallow. Elsewhere there is a passage upwards with no marked change from Cambrian slates or shales into similar slates or shales of Arenig age. Thus in the Lake District the Skiddaw Slates are partly of the one age, partly of the other. Almost throughout Ordovician times the rocks deposited in the British area fall into one or other of two groups—the deep water or shaley facies in which the typical fossils are graptolites, and the shallow water or shelly facies in which shells (especially of brachiopods) and trilobites are typical fossils. Correlation between the two facies is often difficult and where the sequence is almost entirely of the deep water type, as it is in the much folded rocks of the Southern Uplands, separate names were usually given to the several horizons. The shaley facies is found in Wales but it is there that the shelly facies is best developed. In general it would seem that the depth of the water increased from Wales towards the Lake District and Scotland—towards the centre of the old trough.

Silurian times succeeded Ordovician with little change of conditions. The old sea trough suffered renewed buckling and was gradually becoming filled in with sediments so that in Silurian times shallow water deposits were the rule. But there were no great land masses near the British Isles to yield coarse sediment: often conditions of still clear water prevailed which favoured the deposition of shallow water limestones, such as the famous Wenlock limestone which makes up Wenlock Edge in Shropshire and the interesting Wren's Nest at Dudley, as well as the almost equally famous Aymestry Limestone. In contrast to Ordovician times volcanic activity had practically ceased.

Though various divisions of the Silurian have been proposed the old three-fold one into:

(3) Ludlow Series or Ludlovian;
(2) Wenlock Series or Wenlockian;
(1) Llandovery Series or Valentian;

has stood the test of time and is most convenient.

The end of Ordovician times was marked by gentle folding so that the sediments were thrown into a series of broad folds. One well-marked trough remained—maintaining the south-west to north-east alignment of the previous trough—and in the centre of this synclinal

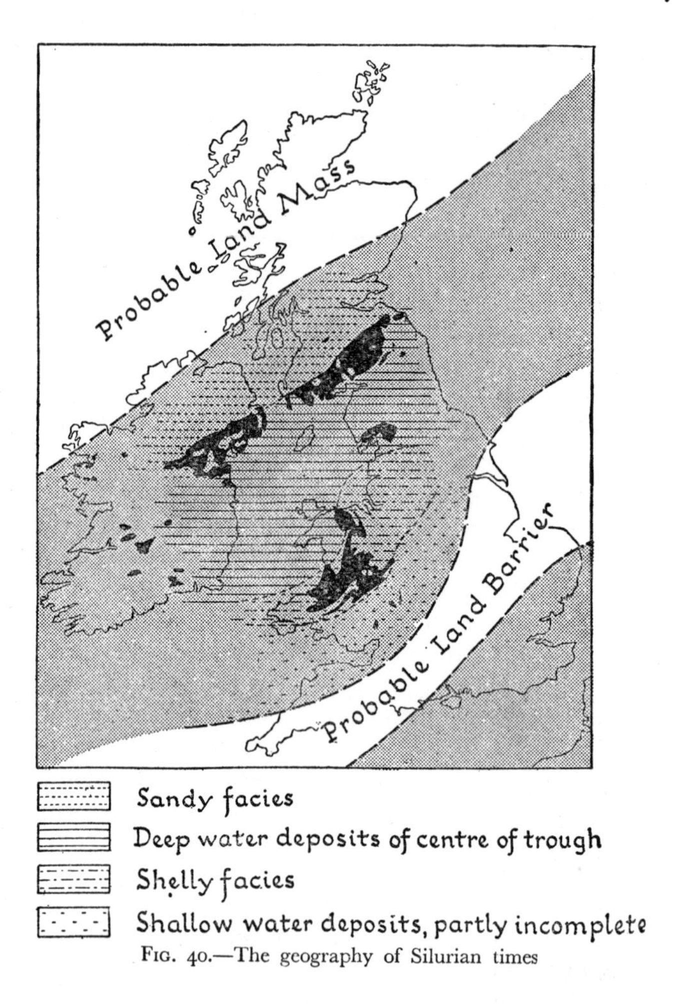

FIG. 40.—The geography of Silurian times

basin the earliest Valentian deposits follow the Ordovician conformably and the sequence is complete. On the margins of the trough there is an unconformity, the older Valentian beds are absent and the younger overlap them. A definite land ridge was formed across England, though its exact position can only be inferred since it is buried under later rocks. In Valentian times there was still a distinction between the graptolitic and shelly facies but the graptolites, approaching their last days, were forced to live in shallower waters while the margins of the trough were marked by the growth of coral reefs. The rocks, both the Wenlockian and Ludlovian, are well seen in Shropshire and are there divisible into :

Upper Ludlow Shales ;
Aymestry Limestone ;
Lower Ludlow Shales ;
Wenlock Limestone ;
Wenlock Shales ;
Woolhope Limestone.

The Woolhope Limestone is interesting since it is built up to a large extent by the remains of a calcareous alga (*Solenopora*) whereas the Wenlock Limestone is largely of corals in the position of growth together with many brachiopods, trilobites and crinoids. The Aymestry Limestone is more local whilst in North Wales and the Lake District the whole Wenlock-Ludlow sequence is of monotonous sandstones and flags, generally becoming coarser in the higher beds.

Silurian rocks are known from deep borings in south-eastern England but were probably laid down in a separate trough, one linked with Brittany and the Ardennes.

The Silurian period was brought to an end by the onset of the main folding movements of the Siluro-Devonian or Caledonian orogenesis. For the most part the folds followed the same general trends as the minor puckerings of the old sea trough, though it is claimed that in some areas angles of 15° or 20° separate the axes of the two sets of folds. The folding reached its greatest intensity in the area of the Scottish Highlands where there are great thrusts and overfolds with " nappes " torn off from their roots and driven far to the north of their original position (see p. 234). The rocks were intensely metamorphosed, though the belief that the rocks of the Grampians represent highly metamorphosed Silurian sediments is now abandoned in favour of an earlier age for them. The Highland sequence represents in fact the base of the greatest of the Caledonides or Caledonian mountains and some of the great granite masses now exposed must have been intruded into the once deep-seated core of the now greatly reduced mountains.

Though it has undergone renewed subsidence at intervals since, the great Rift Valley (see p. 228) of the Scottish Lowlands was initiated at this time whilst the deep water Ordovician and Silurian shales from the centre of the trough were intensely plicated into tiny concertina-like folds to form the Southern Uplands. In the Lake District and in Wales the folding was much less intense—the areas were further from the main epicentre of folding, whilst the thick masses of interbedded lava and intrusive sills must have resisted the minute and intense plication though they underwent folding, faulting and fracturing.

It has already been noted that the famous Old Red Sandstone rocks were deposited in the great tectonic valleys between the main mountain ranges whilst a sea covered the area that is now Devon and Cornwall. In the centres of the basins, notably the Welsh borderland where the Old Red Sandstone was probably laid down in a shallow saline gulf rather than in a basin entirely cut off from the sea, the Silurian rocks pass up almost without a break into the lower beds of the Old Red Sandstone. We have already referred to the interesting bed in this area which most geologists regard as the base of the Old Red, the famous Ludlow Bone Bed, consisting almost entirely of the spines and bony ornaments which adorned the leathery skins of primitive fish-like creatures. One cannot help conjuring up the vision of a group of creatures seeking to adapt themselves to the rapidly changing conditions but being overwhelmed in myriads by a sudden change in salinity of water. It is interesting that fish are so extremely rare as to be virtually unknown in beds older than the Ludlow Bone Bed, whereas in later beds they become the dominant life forms. Many of the curious Devonian " fish "—actually Ostracoderms or lampreys with hard external armour, such as *Pteraspis*—seem to have been able to live both in the Devonian Seas and in the Old Red Sandstone lakes. Land plants first became important in Devonian times and the primitive *Rhynia* is specially noteworthy because specimens with the internal structure well preserved have been obtained from the Old Red of Rhynie in Aberdeenshire.

Where the Palaeozoic Platform is thickly covered with later sediments over south-eastern England, the reconstruction of Devonian geography is difficult, but elsewhere the picture presented in Fig. 41 can be regarded as far more definite than for any preceding period.

During the period there were intermittent earth-movements, so that, for example, the Upper Old Red of South Wales is ushered in by

Plate XV

FINGAL'S CAVE, ISLE OF STAFFA L.M.S. RAILWAY

A good example of hexagonal jointing in basalt—the lava from an old volcano

H.M. GEOLOGICAL SURVEY

DUNCRYNE, GARTOCHARN, DUMBARTONSHIRE

A good example of crag-and-tail. The crag is formed by a resistant mass of lava, an old volcanic plug, from around which less resistant rock was removed by the glacier known as the Loch Lomond Glacier

ARTHUR'S SEAT, EDINBURGH S. H. BEAVER

In the middle distance Salisbury Craigs form a fine example of a volcanic sill. Arthur's Seat, in the background, is a composite volcanic neck

GEOGRAPHICAL PUBNS.

STRIDING EDGE, HELVELLYN, LAKE DISTRICT

A good example of an arête in glaciated country

thick beds of rolled pebbles, but broadly speaking towards the end of the period the mountain masses had been greatly reduced. The Old Red Sandstone is well named in so far as the adjective "red" is

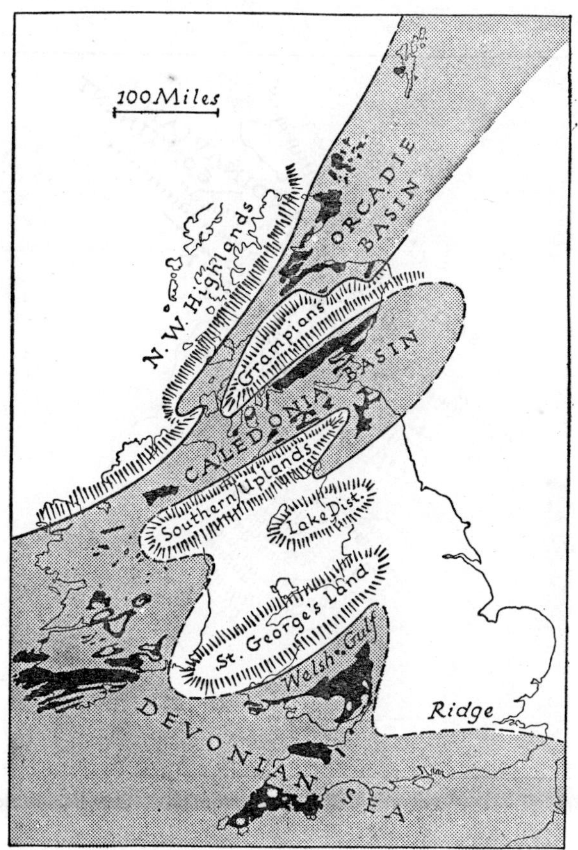

FIG. 41.—Caledonian folding and the geography of Old Red Sandstone times

concerned. Though in the lower part in the Welsh borderland green sandy shales or " marls " occur, they weather red and afford a red soil. Similarly the sandstones and flagstones may be whitish, grey, mottled or orange but the dominant soils nearly always have a reddish tinge and are a constant reminder of the desert conditions which are believed to have prevailed round the basins of accumulation. The inclusion of the word " sandstone " in the title is less fortunate. In some areas—e.g. Herefordshire—the beds of sandstone are definitely subordinate to the immensely thick masses of red marl which break down into

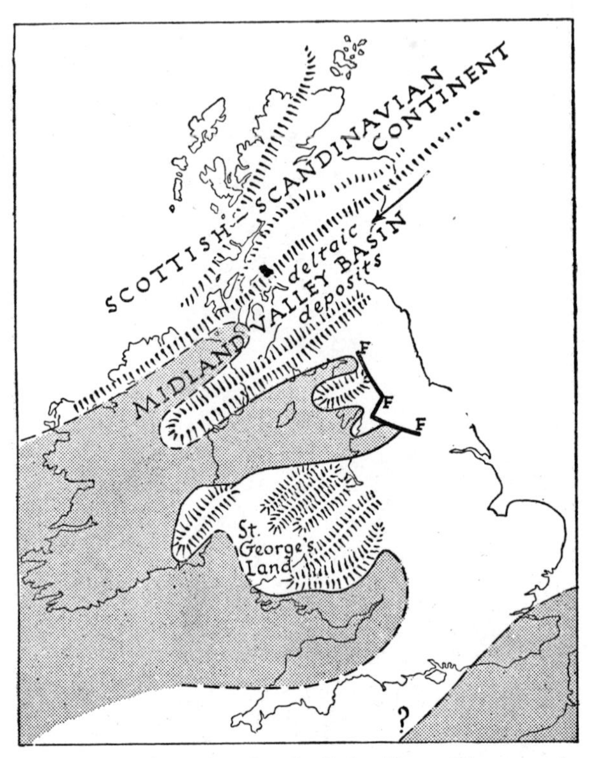

FIG. 42.—The geography of early Carboniferous Limestone times

quite heavy clay soils. Though called " marls " they are non-calcareous. At some horizons in the marl, however, there are bands of concretionary limestone called cornstones and where these occur they help greatly to improve the resulting soil. There is, in fact, the widest possible variation in scenery, landscape and land utilisation associated with the Old Red Sandstone.

The Devonian beds of the South-Western Peninsula, mainly marine, include shales, sandstone (sometimes massive) and locally limestones representing former coral reefs. The significance of these is discussed in the chapter on the area.

Towards the close of Devonian times one may picture the Cale-

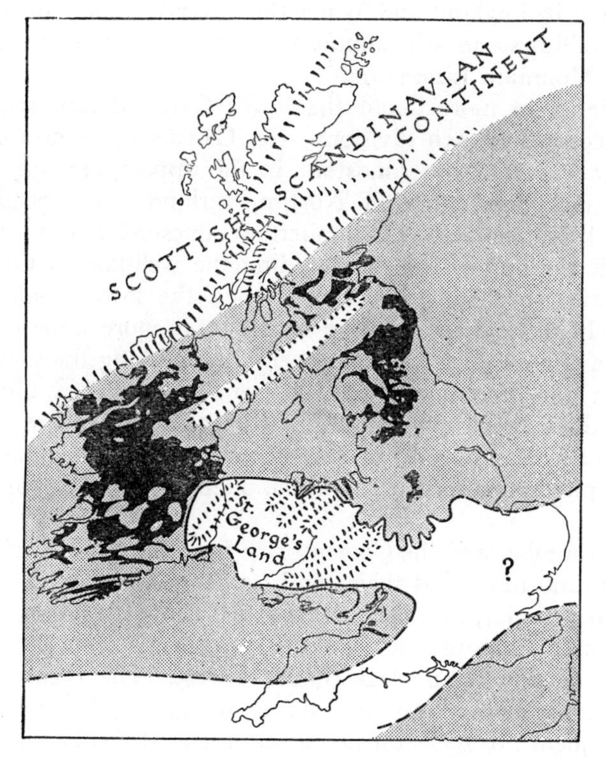

FIG. 43.—The geography of late Carboniferous Limestone times

donian mountains of England and Wales worn down almost to base level and yielding very little sediment whereas the intervening basins had been largely filled in. In the area corresponding with what is now the Highlands of Scotland, a land mass of continental size remained and was far from being worn down to base level. Such was the general landscape which was subjected to a gradual and steady lowering, with the consequent invasion of the basin areas by the Carboniferous sea.

The continental mass of the north yielded much sediment so that the lower Carboniferous deposits in the Central Lowlands of Scotland are sandstones, shales, cementstones and siltstones with only occasional bands of limestone. Traced southwards into Northumberland, limestones gradually assume a greater importance whilst in the main areas of deposition in England and Wales the lower Carboniferous beds are essentially of limestone—the famous Carboniferous Limestone, formerly called the Mountain Limestone.

Considered in more detail the lower Carboniferous sequence in Scotland consists of two divisions—the Calciferous Sandstone Series in the east with a coarse sandstone in the upper part corresponding with the Fell Sandstones of Northumberland (mentioned below) succeeded by the so-called Carboniferous Limestone Series. The latter begin with a group of shales, including the well-known Oilshales of the Lothians, then some limestones, then the lower coal group of Scotland (Edge-Coal Group) and finally some more limestones. The occurrence of coal suggests swampy conditions along the shores of the continent ; the limestone bands represent incursions by the sea. In Scotland there were contemporary volcanoes and there are large stretches of volcanic rock of lower Carboniferous age as well as numerous plugs or infilled necks which often form conspicuous isolated hills.

In Northumberland the Cementstone Series or lower Carboniferous consists of sandstones and shales with some seams of magnesian limestone (cementstone), and then the massive Fell Sandstones, worthy of special mention because they give rise to several important features including the fine scarp of Rothbury Forest. The rocks above the Fell Sandstones include clays and shales, with a few thin, poor coals.

The sequence of rocks on the western side of the Pennines is very different and although the uplift of the Pennines took place much later there is some evidence that faulting caused a separation of the two basins of deposition. The Carboniferous Limestone in this Irish Sea Basin of deposition has been carefully zoned and the evidence is clear that the earlier zones are absent round the margins of the basin but that continued depression of the land resulted in a steady overlap of the sea. As a result the higher limestones, of the *Dibunophyllum* zone, are found over the widest area and on the margins of the old basins rest directly on older rocks of different ages.

In the south-western basin of deposition the position is different—here it is the earlier limestones which have the greater extent. The land seems to have been subject to a " tilt " which threw up the land in South Wales but caused an extension of sea towards the south.

Despite the wide extent of Carboniferous Limestone shown on the geological map of Ireland it is so much hidden by glacial drift that it is difficult to study and much less is known of the geography. It is also difficult to reconstruct the conditions as they were in south-eastern England : the Carboniferous Limestone is well developed in Belgium but in England is hidden by younger rocks though known in borings and colliery shafts in east Kent.

The Carboniferous Limestone is not only of great interest geologically ; it plays a part in determining the surface features of Britain out of all proportion to its actual outcrop. The distinctive landscape features so readily called to mind by the Limestone Pennines, Malham Tarn, Dovedale, the Great Orme, the Mendips with the Cheddar Gorge, the " Rim " of the South Wales Coalfield, and the Avon Gorge are but some examples of the scenic heritage we owe to the Carboniferous Limestone, whilst the effect on flora or land use is no less marked.

Contemporary volcanic activity was not as prevalent in England as in Scotland but did occur—the " toadstones " of Derbyshire are in fact basaltic lavas.

In the upper horizons limestone often gives place to dark shales with sandstone bands and sometimes with thin, black limestones. This is the Pendleside or Yoredale phase, and sediments of this type occur also in North Devon where, in fact, only these equivalents of the upper limestones are found.

The middle division of the Carboniferous is the remarkable group, mainly of massive sandstones but sometimes with interbedded shales, known as the Millstone Grit. It was in the West Riding of Yorkshire that certain beds were long worked to provide grindstones in the early days of the cutlery industry and the working of beds here and elsewhere to provide grindstones for flour mills gave the whole group its name. As a stone suitable for grinding corn it is not difficult to visualise a coarse tough sandstone, somewhat porous and with the grains rough and angular and such a description would fit many of the beds of the Millstone Grit. Clearly these beds are in marked contrast to the massive limestones which are so typical of the lower horizons of the Carboniferous sequence and indicate a marked change in geography.

There seems to have been an uplift in the continental mass which lay over Scotland and to the north and the Millstone Grits are of the nature of great deltaic deposits brought down by huge rivers draining from the north. The courses of the main river, its tributaries and distributaries (as the delta streams are called) shown on Fig. 44 are

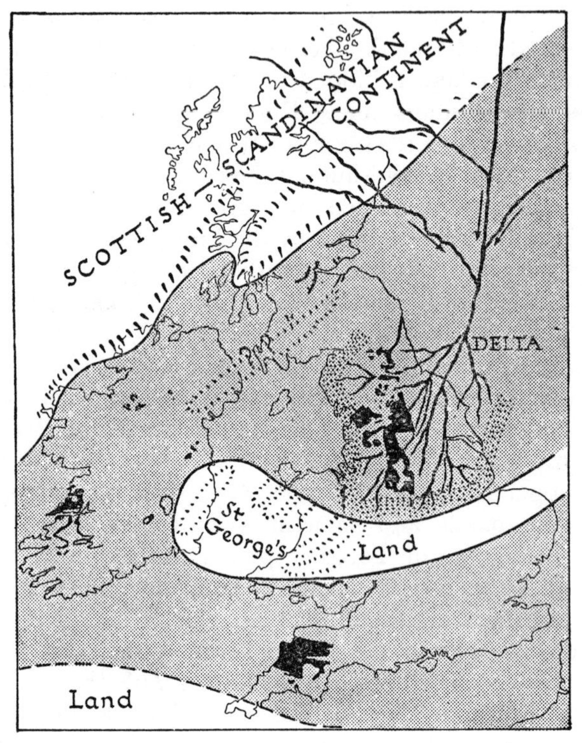

Fig. 44.—The geography of Millstone Grit times

not purely imaginary, but are based on the fascinating study of pebbles in the grits which can be matched with parent rocks to be found in situ not only in different parts of the Highlands of Scotland but, in the case of certain very distinctive types, in Norway. As the latter country provides the only known occurrences of the rocks in question, the rivers must either have drained from there, or there must have been other occurrences of the rocks in the continental area between Scotland and Scandinavia since worn away.

Naturally the Millstone Grit deposits vary enormously in thickness. Where they are especially well developed in the Peak District of Derbyshire they attain some 4000 feet—four horizons of sandstone known respectively as the Fourth or Kinderscout Grit, the Third Grit, the Second Grit and the Rough Rock or First Grit (the latter being at the top of the series) separated by shaley horizons.

The Millstone Grit affords some evidence of the conditions prevailing on the land mass from which the sediments were derived. The coarseness of the sediment points to high elevation of the land with strong mountain torrents, the freshness of the fragments of felspar amongst the sand grains—an unusual mineral to be found as sand grains because it is unstable and easily destroyed by long transport—suggests the work of insolation or the sun's action on a rugged bare land. At the same time such a high land mass, heated by the sun's rays, would cause a monsoon, the winds blowing from the south or south-west, just as one finds in southern Asia at the present day, and this helps to explain the luxuriant Coal Measure vegetation which flourished shortly afterwards on the lower ground.

The so-called Millstone Grit of more southern areas seems to be mainly of the nature of a shallow water or shore deposit which accumulated round the island of St. George's Land (see Figs. 41-44).

The Millstone Grit deltas and the shore deposits round the continental coasts or islands—in other words the silting up of the Carboniferous Limestone seas—prepared the way for the growth of swampy forests which characterise the Coal Measure period. At different times these swamp forests flourished over the greater part of England and it was Lyell who, a century ago after a visit to the Great Dismal Swamp of Virginia and North Carolina, drew the parallel between that present-day area and what he believed Coal Measure swamp forests to have been like. There are in fact many areas of deltaic swamp forests, including those of the Amazon, which provide comparable modern examples. A reconstruction of a Coal Measure forest is shown in Plate 15A whilst the disposition of coal seams in a typical British field is shown in Plate 12—the West Cumberland field where coal-seams are worked under the Irish Sea.

At different times and in different areas the Coal Measure forests were overwhelmed by sediment brought down by rivers and even slight changes in relative levels of land and sea would have been sufficient to cause great changes. Actually subsidence seems to have gone on periodically so that new forests sprang up only to be overwhelmed in due course by fresh masses of sediment. Sometimes the subsidence was local so that the growth of the forest was continuous in one area—giving a thick seam of coal—whilst a short distance away subsidence and the deposition of sediment interrupted the growth of the forest so that two or more seams of coal occur. This is actually the case with the famous Thick Coal of the Black Country which is formed from a forest which grew on the margins of St. George's Land but which gives place northward to a number of separate seams. Lithologically it should be noted that the Coal Measures consist of great thicknesses of sandstones and shales in which the coal seams themselves are but minor episodes.

At other times in the Coal Measure period slight but general subsidence took place so that the sea flowed in over large areas and left its trace in the form of thin " marine bands " especially valuable as datum lines in correlation. The source of the material for the Coal Measure sandstones has not been examined in the same detail as it has for the Millstone Grit, but in general was probably the same—again the chief land mass was to the north and it is probable that St. George's Land remained throughout above sea level.

The exact origin of coal has long been the subject of much patient investigation as well as of hot dispute. In general each seam is underlain by a greyish " fire-clay " (so called because it is often used to make refractory or fire bricks) which is believed to represent the muddy ooze in which the swamp forests grew. Though remains of roots are often found in it, it is still possible that the greater part of the material which formed the parent substance of the overlying coal seam may have been drifted into position. The rival " in situ " and " drift " theories doubtless each contain an element of the truth and it is generally agreed that drifted material may sometimes bulk largely in supplying the parent mass. In any case it is agreed that the parent material of coal is vegetable matter. The old idea that the vegetable material changed first into peat, then into brown coal or lignite, then into bituminous or ordinary household coal, and finally into anthracite is no longer regarded as tenable. Although this order is roughly one of decreasing volatile constituents—which might well be eliminated in the course of geological time and with folding and metamorphism—it is now agreed that there must have been fundamental differences in parent material. Further it is clear that the change from the mother substance to the hard coal much as it appears to-day must have taken place almost at once (geologically speaking) for slight folding movements brought seams of coal under the action of denudation even during Coal Measure times and the coal was already sufficiently hard to be rolled into coal-pebbles which are found in later beds. The change from the mother substance to coal probably took place under the influence of certain bacteria. Not only were coal seams subjected in this way to almost contemporaneous denudation but a change in the course of delta streams often resulted in channels being cut in seams recently formed. These channels were later filled with sediments and constitute the "wash-outs" well known to miners. Contemporaneous denudation gave rise also to the formation of clay pebbles and so to claystone conglomerates.

When a coal forest was overwhelmed it was sometimes by sand, sometimes by mud, and so the deposits which form the roofs of coal seams vary greatly. All miners know the value of a hard roof of consolidated sandstone.

Whilst it is easy to picture the deltaic swamp conditions under which the Coal Measure forests flourished and although we know that similar conditions occurred at other geological times and that coal seams of varied age are found in different parts of the world, it remains true that the essential conditions were extraordinarily widespread over the earth's surface at the time of the Coal Measures. There were two great floristic provinces. One is characterised by the *Glossopteris* flora and is found in the various parts of what is believed to have been a continuous land mass (Gondwanaland), though now occupying separated parts of India, Australia and South Africa. The other comprised the whole of the European and North American region. What is remarkable is the relative uniformity of the flora which suggests but little difference in climate over the whole great area. Tree-ferns which, though botanically distinct from the dominant Pteridosperms of Coal Measure times, are the nearest in appearance to the plants of the coal forests, are tropical or subtropical. Thus the general idea has grown up that the climate of Coal Measure times was warm and wet and that such conditions were found as far apart as Spitsbergen and the Antarctic continent in both of which areas coal seams occur. Some botanists contend, on the other hand, that the detailed structure of the Pteridosperms gives no evidence of tropical conditions and in many areas there are traces of almost contemporary glaciers. It may also be noted that some geologists have put forward the idea of an excess of carbon dioxide in the atmosphere as partly explaining the widespread luxuriance of the vegetation.

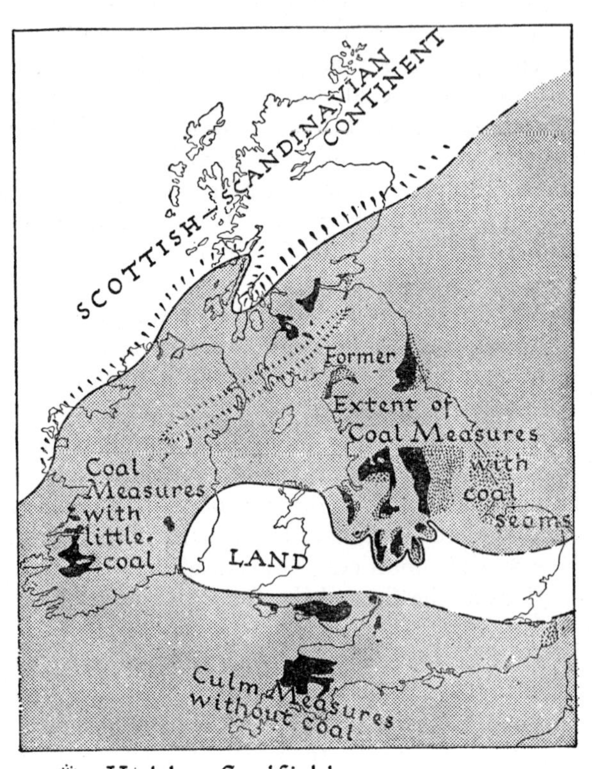

FIG. 45.—The geography of Coal Measure times

Whatever may have been the character of the climate of Coal Measure time, towards the close of the period in England arid, desert conditions began to set in. They set in earlier in the north, and in Scotland there is a great thickness of barren red sandstones and shales on horizons which represent the productive upper beds of the English

Coal Measures, whilst the similar red Etruria Marls and Keele Beds of North Staffordshire are probably of the same age as the higher coal seams of the Radstock Field in Somerset.

Reference has been made to coal seams of different ages, but the zoning of the Coal Measures presents various difficulties. A scheme based primarily on the sequence of marine bands has been devised and so have others based on the sequence of non-marine molluscs and gradually these schemes have been correlated with the sequence based on the study of plant remains. One difficulty remained unrealised until a Welsh mine-foreman and self-trained geologist, as recently as the second decade of the present century, conceived the idea of making huge collections of the plant remains associated with *each* coal seam. He proved conclusively that different plant assemblages were associated with the different seams and that there must have been differences, it may be small, in the habitat conditions. This pioneer study in what may be called palaeo-ecology earned for Mr. David Davies the Honorary Degree of Master of Science of the University of Wales; his early death was a great loss to the ranks of those geological amateurs who have done so much to advance the science.

Broadly speaking it may be said that the Carboniferous Period was brought to a close by the great Armorican earth movements but, as with other great orogenic movements, these began before the end of the period and grew gradually in intensity. One result of the early movements seems to have been the gradual rise of the "Mercian Highlands"—following partly along the line of the old St. George's Land—against which the Coal Measures of south Staffordshire are banked. Concurrently there was a long slow movement of depression to the north, so that the greatest thickness of Coal Measures occurs in an east-west belt passing through Lancashire. It would seem that, although one effect of the Armorican folding was to accentuate or revivify older folds, the characteristic east-west direction of truly Armorican folds was established early. The existence of a resistant land-block or massif in north and central Wales seems in part responsible for the establishment also of north-south folds, well-exemplified by the tightly folded Malvern ridge and also by the more open Pennine anticline of northern England of the same date.

The Armorican folds—alternatively called Hercynian or Carbo-Permian or Permo-Carboniferous—threw the Coal Measures into a series of basins and separating anticlinal ridges. From the ridges the

Coal Measures were rapidly worn away by denudation, leaving the coal basins nearly as they are at the present day. The Armorican mountains and valleys determined the main pattern of the geography which was destined to persist throughout the Mesozoic till changed by the Alpine movements of the mid-Tertiary.

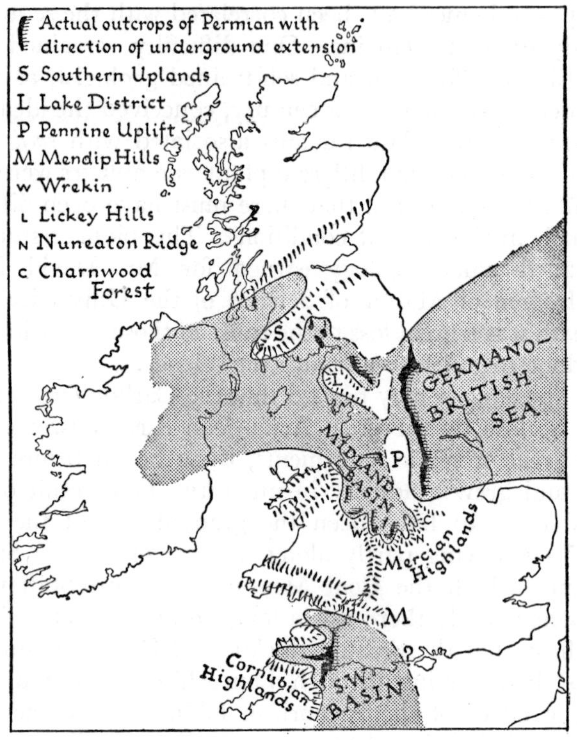

FIG. 46.—Armorican folding and the geography of Permian times

In early Permian times conditions repeated those found in early Old Red Sandstone times. The older geologists who used the term New Red Sandstone for the red beds of the Permian and Trias introduced this very useful comparison and there has been a move in recent years to go back to that term. The Permian basins are shown in Fig. 46. In them the earlier deposits were of the nature of rock screes, with coarse pebble beds very like those one may see at the present day being swept down by mountain torrents from the Himalayas. The Armorican folding isolated a great arm of the sea which covered much of what is now Germany and the North Sea and which stretched also over northern England east of the Pennines. In this

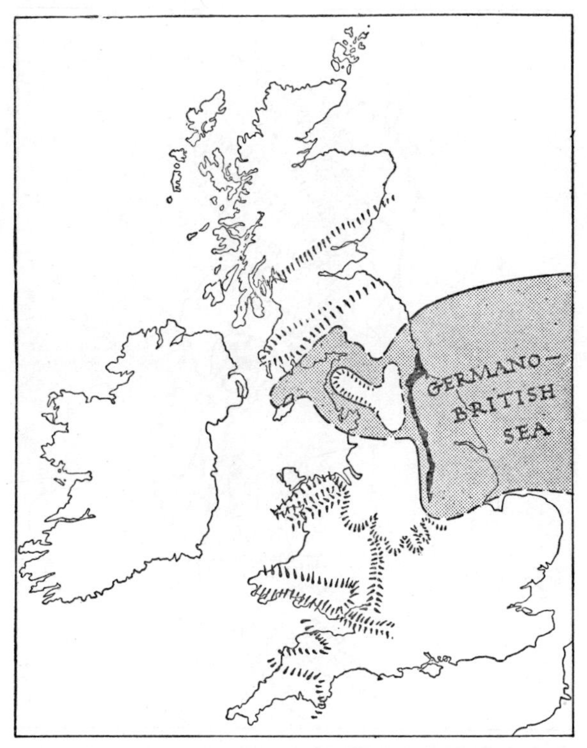

FIG. 47.—The geography of Magnesian Limestone times

great " Dead Sea " salt-bearing red marls were laid down, with a stunted marine fauna (compare the stunted molluscs of the very salt Red Sea of the present day) followed by the interesting dolomitic limestone or Magnesian Limestone so well seen in parts of Durham and Yorkshire.

The sparsely fossiliferous Permian rocks of Britain give little idea of the life of the period. The basins here were probably largely desert in character : elsewhere large reptiles roamed the land and the plant forms were a natural development of the Coal Measure flora, showing few important differences.

Although the Permian is grouped as the youngest of the Palaeozoic

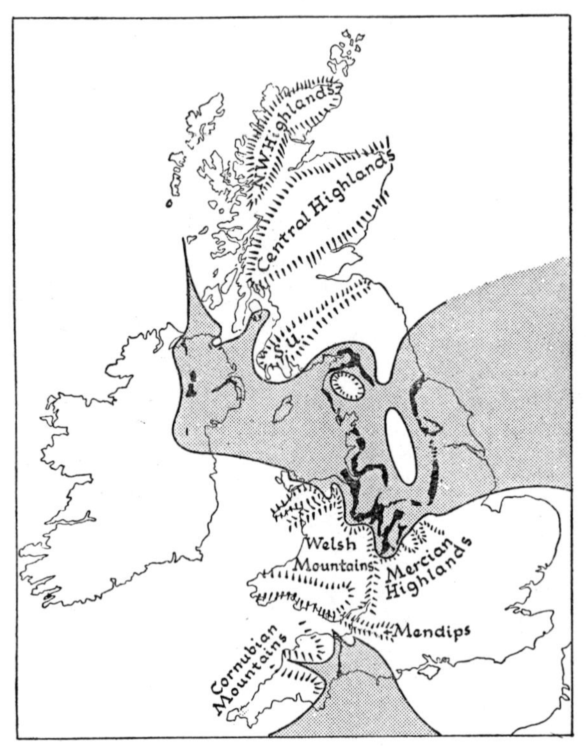

FIG. 48.—The geography of Bunter Sandstone times

periods, there is no break between its deposits and those of the succeeding Trias. In Britain the limestone horizon of the Muschelkalk which lies between the lower (Bunter) beds of the Triassic and the upper (Keuper) is absent. In Bunter times there were in Britain two main areas of deposition—the old South-Western Basin of Permian times and the larger northern basin over the Midlands and north of England

and stretching into Scotland and Ireland. The Bunter beds are predominantly red sandstones, or sands, often current-bedded, and pebble beds. The latter include the famous Budleigh Salterton Pebble Beds of the South-Western Basin and the Bunter Pebble Beds of the Midlands which occur between the Lower and Upper Mottled Sandstones. The

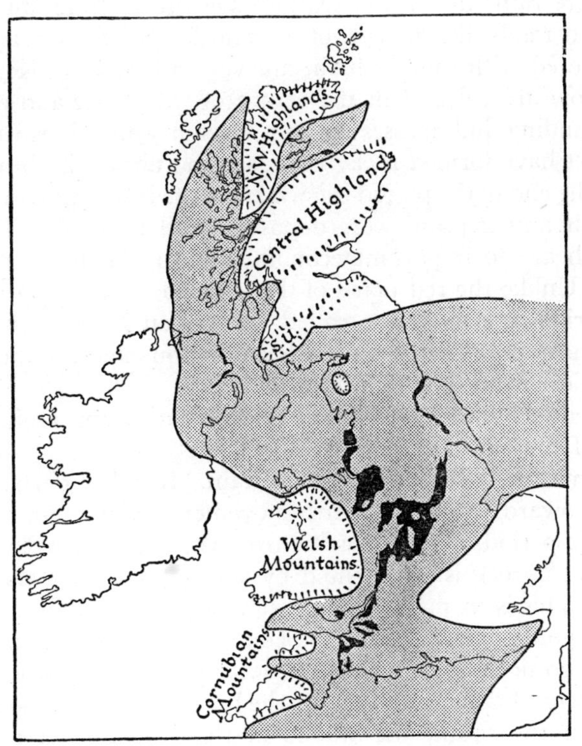

FIG. 49.—The geography of Keuper times

pebble beds are remarkable for the large size of the pebbles and their well-rounded form. In the Midlands the pebbles are largely of quartzites probably derived from Old Red Sandstone quartzites of Scotland so that they appear to have been swept across the desert basin by torrents from the north. Keuper times seem to have been marked by a lowering of the surface, or it may be that the filling up of the basins with sediments caused a wider spread of the Keuper lakes. In any case the Keuper deposits, which include the Keuper Waterstones at the base, overlap the Bunter and are of wider extent. A careful study of the " internal evidence " afforded by these micaceous, brownish current-bedded or even-bedded sandstones suggests that they were deposited in a shallow, brackish-water gulf bordered by sand dunes where rains in the wet season gave rise to transient pools with sun-cracked muds like the *vleis* of present-day South Africa and which were bordered with quite a luxuriant vegetation. The Keuper Marls which follow are thickest in the centre of the basins and wrap round such upstanding hill masses as the Wrekin and Charnwood Forest which must have formed rocky islands rising above the brackish lake. Towards the end of the period the waters of the lake began to evaporate: beds of salt and gypsum were deposited and it is these Triassic salt beds which are so important economically in Cheshire and again in Durham. Unlike the red marls of the lower beds, the upper marls are often a curious green—the so-called Tea Green Marls—with bands of whitish sandstone. The green marls oxidise on exposure and yield red soils like those from the other Triassic beds.

The gentle foundering of the Triassic landscape amid the quiet invasion of the salt-lake flats by the Rhaetic sea ushered in a long period of marine sedimentation in Britain. In other parts of Europe, especially towards the south, earth movements were more marked and the Rhaetic period, taking its name from the Rhaetic or Rhaetian Alps (part of the Tyrol), is represented by very considerable thicknesses of sediment. This is in marked contrast with Britain where, though the Rhaetic deposits extend across the country from Devonshire to Yorkshire, they do not exceed a hundred feet in thickness. It is interesting to picture what happened when, by the breaking down of some barrier, probably in the south, the waters of the Rhaetic sea invaded the Keuper lake. The water would naturally flow, perhaps quite gently but rapidly and surely, over practically the whole of the old lake flat. One can picture vast numbers of Triassic reptiles being overwhelmed by the inrush of the sea. Their remains would become mingled with those of fish which had been living in the salt lake and with those of Rhaetic fish which, swept in with the sea, were unable to survive the conditions of the stagnant salt-water of the lake. Hence there is frequently, very near the base of the Rhaetic, an important " bone bed." The hardier of the marine molluscs brought in by the sea

L. D. STAMP

Blaen-Rhondda Colliery at the head of the Rhondda Valley, South Wales (See page 219)

PLATE 14

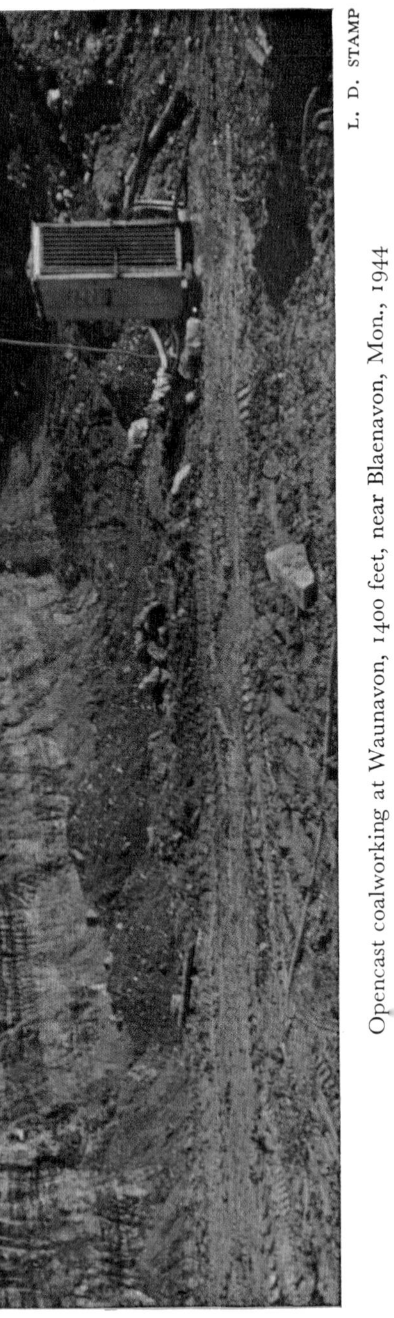

L. D. STAMP

Opencast coalworking at Waunavon, 1400 feet, near Blaenavon, Mon., 1944

survived for many generations, but only in stunted forms. In some places molluscs are found in beds below the Bone Bed in greenish or grey marls exactly like, except for their presence, the underlying Tea Green Marls of the Keuper, and this suggests temporary incursions of the sea before the main invasion. Throughout the Rhaetic period the surrounding lands were low and yielded little sediment so that the Rhaetic beds, though thin, represent a long period of time. They comprise dark shales, marls and pale-coloured limestones. One band of limestone, the fascinating Cotham marble, was often seen polished in fragments on Victorian mantlepieces, for it has curious moss-like or arborescent markings in a creamy background. It has been suggested that the markings were formed through the agency of bubbles of gas, originating in the lower layers of a soft creamy deposit, rising into higher layers and there bursting and scattering minute specks of darker material. In places the bed appears to have been broken up, probably by disturbing currents, and recemented as in the "False Cotham" of Aust Cliff on the River Severn. The Rhaetic deposits, especially the limestones known as the White Lias, may give rise to a low "scarp"—the lowest of the scarplands which are such a marked feature of the South-east Midlands. The Rhaetic is interesting too for the remains it has yielded of the earliest British mammal—*Microlestes*. *Microlestes* was a small creature, only about the size of a rat.

The poverty of the Rhaetic fauna in Britain and the small thickness of the beds may indicate that the marine incursion was but a temporary one. In any case it can be said that the Liassic (Lower Jurassic) witnessed either a renewed marine invasion of the old Triassic basins or a continuance of the Rhaetic invasion—the Liassic sea was roughly co-terminous with the fullest extent of the Keuper lakes.

Jurassic rocks play a very large part in influencing the scenery of Lowland Britain. It is not only that they consist of an alternating series of clays and limestones or sandstones which are respectively weak and strong in their resistance to atmospheric weathering and so give rise to all the varied forms of scarpland previously described (see pp. 67-68) according to the amount and direction of the local dip, but in detail the sequence varies greatly from locality to locality. A limestone, thick and important in one area and forming a noble scarp such as that of the Cotswolds, may die away so that both limestone and resulting scarp may be absent from that horizon in another area. These local variations in rock sequence are more noticeable in the

Middle and Upper Jurassic than in the Lower Jurassic or Liassic.

The Liassic beds follow naturally and without a break those of the Rhaetic and, were it not for the very different position on the continent of Europe, there would in England be no reason for separating the " White Lias " or Rhaetic from the " Blue Lias " or " Hydraulic

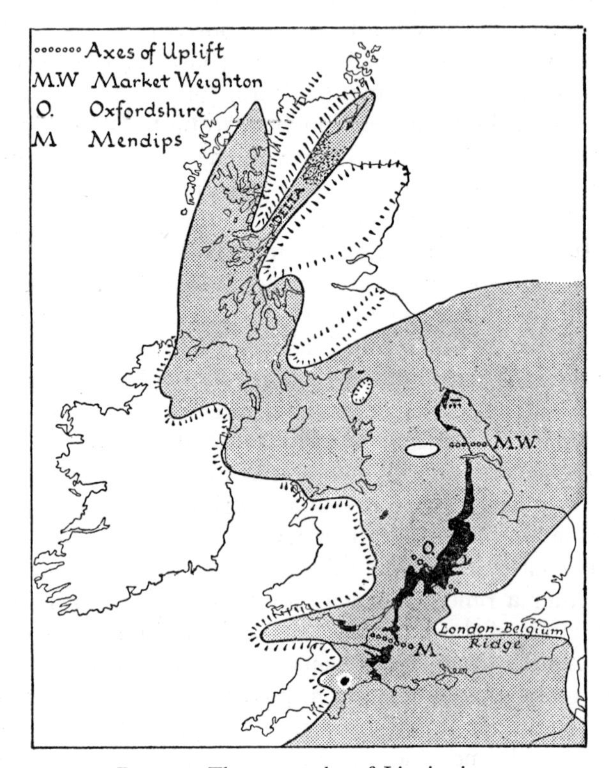

FIG. 50.—The geography of Liassic times

Limestones" of the lower true Lias. These lowest Liassic Beds consist of alternating greyish clayey limestone beds, from a few inches to a foot or so in thickness, and clays. The limestones have the proportion of clay to lime which permits them being calcined direct to form hydraulic cement (cement which sets under water) and hence are quarried at various localities. The remainder of the Lower Lias is mainly of clays which give rise to a broad plain in the Midlands of England, scarcely separable from the Triassic plain except by colour of soil. Only sometimes do the Rhaetic beds or the Blue Lias limestones give rise to a small ridge or scarp separating the Triassic and Liassic plains. In some areas (notably the Clevelands, Lincolnshire and Oxfordshire) the Middle Lias includes important ironstones; elsewhere the "Marlstone," found at this horizon, though not valuable as an iron ore, gives rise to a prominent scarp. In still other areas the Middle Lias is of sands or sandy clays whilst the Upper Lias, frequently relatively thin, is generally of clays passing up into soft sands. The low ground developed on Liassic clays frequently affords but dull scenery and where these beds meet the sea muddy foreshores detract from coastal amenities. To any one interested in geology, however, the varied ammonites of Liassic beds are amongst the most fascinating of fossils.

In places, notably in South Wales, the normal Liassic deposits can actually be traced laterally into shore deposits, whilst in the Moray Firth area of Scotland deltaic deposits of Liassic age are found.

Traced across England from the Dorset coast to Yorkshire the beds of the Lias are found to vary greatly in thickness. A scheme of zones, based mainly on characteristic ammonites, was proposed as long ago as 1856 to 1858 by Oppel and though many refinements of his scheme and a great elaboration of sub-zones and hemerae has since been proposed, the amateur can still recognise the fossils—many of them now regarded as groups of species—which Oppel proposed to typify each of his zones. It is found that the sequence of zones is complete, or almost complete, even where the whole thickness of beds is very small. This suggests that there were several submarine ridges or axes along which there was little, though continuous, sedimentation, whereas the intervening basins underwent slow and gentle but almost continuous subsidence which permitted the deposition of a much greater thickness of sediment. There were, in fact, three such axes, separating four main basins of deposition. The first axis was the old Mendip uplift—an Armorican line with a trend from slightly north of west to south of east. This separated a southern or Dorset-Somerset basin of deposition from that of Gloucestershire or the Cotswolds. The next axis was not quite so marked or persistent: it may be called the Oxfordshire axis and to the north of it lay the great basin of Northamptonshire and

Lincolnshire. The third axis is also a well-known one; again it followed a roughly east-west or Armorican trend and is called the Market Weighton axis from the Yorkshire town of that name. To the north of it lay the north Yorkshire basin in which the sequence of deposits is often very different from that further south.

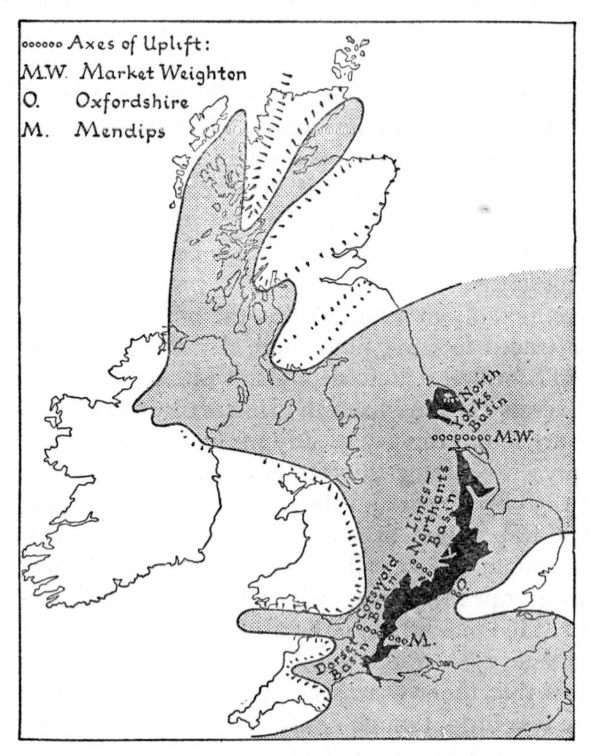

FIG. 51.—The geography of Middle Jurassic times

These axes were well marked in Liassic times and persisted into Middle and partly into Upper Jurassic times. Actually their influence was particularly marked in Middle Jurassic times when there were great differences in the deposits laid down in each basin. The surrounding low land surfaces yielded little sediment and in the clear though shallow waters limestones were formed at intervals, though at other times the sea waters were disturbed by cross-currents which resulted in current-bedded sandstones, whilst some of the detrital limestones are also current-bedded. The important effect of certain of the limestones and sandstones on present-day surface features is considered more fully in the chapter on the Scarplands.

Middle Jurassic times were marked not only by warping movements which resulted in a deepening of the synclinal basins but also by the development of a general tilt from north to south. Thus the Inferior Oolite or Bajocian beds are represented on the west coast of Scotland by limestones, the succeeding Great Oolite or Bathonian by deltaic deposits of coarse sand. In Yorkshire there are limestone bands but the Lower, Middle and Upper Estuarine beds there represent the Middle Jurassic and tell their own story regarding the proximity of land. At the same time the tilt caused a transgression of the sea against the old ridge under London and North Kent.

The highest bed of the Middle Jurassic is the Cornbrash, so-called because it yields a brashy or stony soil particularly suited to corn growing. This relatively thin limestone indicates the establishment of uniform marine conditions, whilst the succeeding Kellaways Rock, usually regarded as the base of the Upper Jurassic, tells the same story. Then followed a change, the nature of which is less easy to state than the results. It was probably a general slight lowering of the surface, accompanied by a change of currents, so that the mud-laden waters of a dirty sea spread over the whole extent of the old basins, destroying their clear-water coral faunas. The muddy deposits of this sea form the Oxford Clay which persists with but little change right across the country from Dorset to Yorkshire. It is remarkably constant in thickness until one reaches the Market Weighton axis in Yorkshire, against which it thins rapidly, so that one presumes the axis was still rising. There is no sign of the persistence of the Mendip axis but towards the end of Oxford Clay times a new anticlinal ridge can be traced. It may have followed the old line called above (page 131) the Oxfordshire axis, or the still older Charnian line from north-west to south-east across the Midlands, but in any case it effectively cut off the muddy northern waters from the clearer ones in the south. Thus, north of Oxford the Oxford Clay is succeeded by the similar Ampthill Clay; south of Oxford by the limestone known as the Corallian. Where the latter occurs, it usually gives rise to a ridge with loamy corn-growing soils thus interrupting the general stretch of the grassy clay lowlands.

Then once again muddy waters spread over the whole and the deposits formed are known as the Kimeridge Clay (authorities differ as to whether the village on the Dorset coast from which the clay takes its name should be spelt with one " m " or two) which is one of the most persistent of the Jurassic deposits. It varies from over 1000 feet

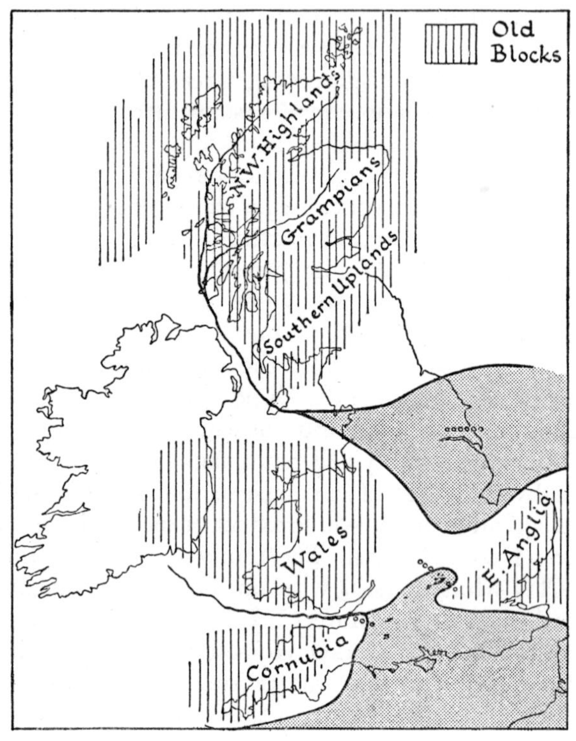

FIG. 52.—The geography of Upper Jurassic times

in thickness in Dorset to less than 100 in Bedfordshire along the Charnian axis, increasing northwards to 500 feet or more, thinning against the Market Weighton ridge and then increasing again. It is a dark grey or black clay, shaley and bituminous especially in the upper part where the sulphurous oil-shale band known as the Kimeridge

Coal occurs in the south. Almost everywhere the Kimeridge Clay coincides with low ground and because of the thickness of the beds and wide extent of the outcrops these plains are often very extensive (see below, p. 197).

At the close of Kimeridge Clay times or a little later the Northern Sea was definitely cut off from the Southern. In the Northern Sea some Portlandian beds were laid down but they were eroded away later and the only remaining traces are some rolled Portlandian fossils, having affinities with Russian species, found at the base of the mainly Cretaceous Speeton Clay. The source of mud having been cut off, the tranquil waters of the Southern Sea provided ideal conditions first for the deposition of some fine calcareous sands and then for the formation of the valuable shelly, oolitic limestone known as Portland Stone. This Southern Portlandian Gulf with its chemical deposition of calcium carbonate to form the grains of the oolite (see p. 200) seems then to have been cut off from the sea, and the succeeding Purbeck Beds consist of alternating estuarine and freshwater beds with marine intercalations. The freshwater beds include the famous Purbeck Marble which consists largely of the shells of the freshwater snail *Paludina*. This marble, a dark rather dull green when polished, was largely employed by mediaeval church builders in small columns and interior decoration as in the Temple Church, London, and Salisbury Cathedral. Here and there are found dirt-beds—actually old soils formed during terrestrial interludes. Later the Purbeck lagoon, quite cut off, became the Wealden Lake in the fringing jungles of which roamed giant reptiles such as *Iguanodon*. These Wealden deposits, though usually regarded as Lower Cretaceous, thus mark the natural end of the major cycle of sedimentation which had persisted from Permian times.

In the meantime, in the north, it is possible that northern England was raised above sea level for a very short period in Portlandian times; in any case, the area came under erosive wave-action. Then the sea returned in late Portland or Purbeck times and the deposition of the Speeton Clay commenced—contemporaneously with the sandy beds of Lincolnshire known as the Spilsby Sandstone.

There is some difficulty in classifying the lower beds of the Cretaceous system in Britain because of the differences in the conditions of deposition in the northern and southern areas. The following table expresses the general sequence :

	Northern Area	*Southern Area*
Upper Cretaceous	Upper Chalk or Senonian	Upper Chalk or Senonian
	Middle Chalk or Turonian	Middle Chalk or Turonian
	Lower Chalk or Cenomanian	Lower Chalk or Cenomanian
	Red Chalk	Gault
Lower Cretaceous	Speeton Clay and	Lower Greensand
		∿∿∿∿∿∿∿∿∿∿
	Spilsby Sandstone	
		Wealden
		Purbeck
		Portland
	∿∿∿∿∿∿∿∿∿∿	
Jurassic	Kimeridge Clay	Kimeridge Clay

Note.—The wavy lines indicate gaps in the sequence or unconformities.

It is not important whether we class the Wealden as Jurassic or follow the ordinary English usage and call it Lower Cretaceous. The important point is that a lake or lagoon, possibly linked with the sea, occupied a large area in south-eastern England at the same time as the Speeton Clays were being deposited in a marine gulf in the north. Then followed a lowering of the land surface with a marine invasion so that a transgression of the sea against the central land ridge took place from two directions—from the north and from the south. In early Cretaceous times, however, the two basins of deposition, though both were marine, remained separate and the faunas are very distinct. The ammonites, for example, which characterise the northern province are mostly species well known in Russia. The dividing ridge of land which thus persisted for so long had a definite trend from north-west to south-east through the region of Charnwood Forest in the direction of London and thence, rather more easterly, to Belgium. It is perhaps an interesting example of the resuscitation of uplift along a very old line of movement—in this case actually a pre-Cambrian or Charnian line. This " posthumous Charnian " folding is a very good example of the dangers of using " Charnian " or " Caledonian " or " Armorican " merely as implying *direction*. Some geologists use Charnoid, Caledonoid or Armoricanoid when they wish to imply direction irrespective of age.

It is probable that temporary connection was established across the ridge in Lower Greensand times but in the central area deposits of this age are mostly sands, probably deposited in erosion hollows, since they vary rapidly in thickness and have a discontinuous outcrop. An interesting example of one of these local deposits is the Sponge Gravel of Faringdon in Berkshire, consisting largely of calcareous sponges but with many rolled Jurassic fossils. A patch of sand at Seend in Wiltshire is interesting because it has been worked as iron ore. Details are given of the characters of the Lower Greensand deposits in the main southern area in the chapter on the Weald, where they play such an important part in the determination of scenery.

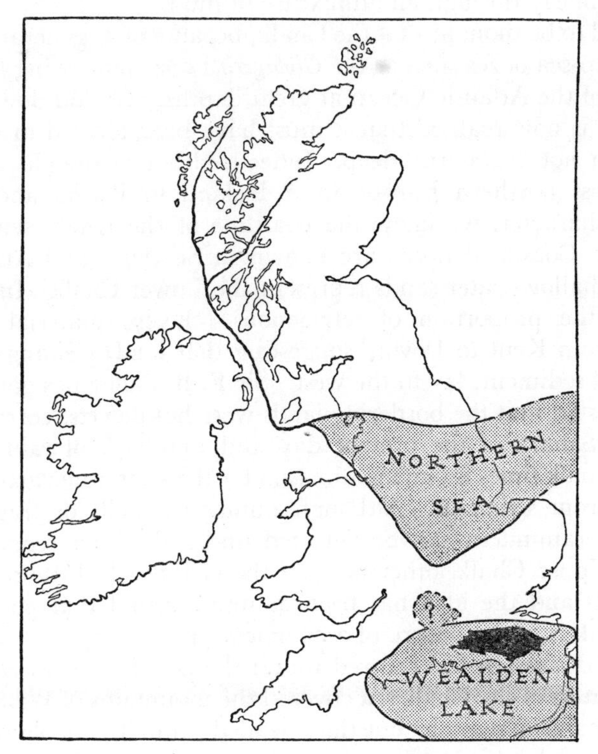

FIG. 53—The geography of Lower Cretaceous times

Permanent connection across the central ridge was probably not established till Lower Chalk times, for the Gault of the south—dark sticky clay—is still very different in lithology and fauna from the Red Chalk of Hunstanton, Lincolnshire, and Yorkshire which is on approximately the same horizon. In Bedfordshire it is the Lower Greensand which rests unconformably on Oxford Clay : under London the Gault rests directly on Palaeozoic rocks and in Belgium the continuation of the ridge was not covered by the sea till late in Chalk times.

Indeed, after Lower Greensand times, the story over north-west Europe is almost everywhere the same—it is that of a great marine transgression. Frequently referred to as the "Cenomanian Transgression," it actually commenced earlier. The lands invaded were relatively low and yielded little sediment—though the Lower Chalk is grey and marly through an admixture of mud.

It used to be thought that the Chalk, because of its general similarity to the deep-sea oozes such as the *Globigerina* ooze now being formed on the floor of the Atlantic Ocean at great depths, was laid down in deep water. It is now realised that it must have been formed in very clear water, but not necessarily deep. Indeed, although the chalk is found right across northern Europe from Ireland to Russia and changes little in character, we know the coastline of the Chalk sea was not far away. Coastal deposits are known in Scotland and Austria ; in Belgium shallow-water sands represent the Lower Chalk. In southern England the proportion of terrigenous (clayey) material increases steadily from Kent to Devon, suggesting that land, yielding a certain amount of sediment, lay to the west. Sir E. B. Bailey has gone further and suggested that the bordering lands were hot deserts, corresponding with the Sahara of the present day and that lack of rain on these lands of Chalk times is one main reason for the sea's abnormal purity or freedom from sediment. Although numerous shells or fragments of shells of foraminifera can be detected under the microscope in some types of Upper Chalk, other parts of the chalk yield little evidence of such fossils and the idea has been gaining ground that part at least of the chalk is of the nature of a chemical precipitate.

Much discussion has centred round the problem as to whether or not the waters of the Chalk sea covered the mountains of Wales. There are no remains of deposits but the so-called summit peneplane suggests marine peneplanation. This is most probably of Cretaceous date though claimed by some geologists as Triassic or Jurassic.

In England there is a marked contrast between the white Upper Chalk and the greenish Tertiary sands or mottled clays which are found in immediate superposition, yet there is little if any discordance in the dip of the respective beds. It would appear that the end of Chalk times was marked by a general retreat of the sea. Actually it began in France even before the transgression had reached its maximum in England. The movement was a gradual one and little basins were left, notably in France and Belgium, in which chalky strata continued to accumulate. These strata form, in fact, transition beds between the Cretaceous and the Tertiary and are sometimes grouped with the one,

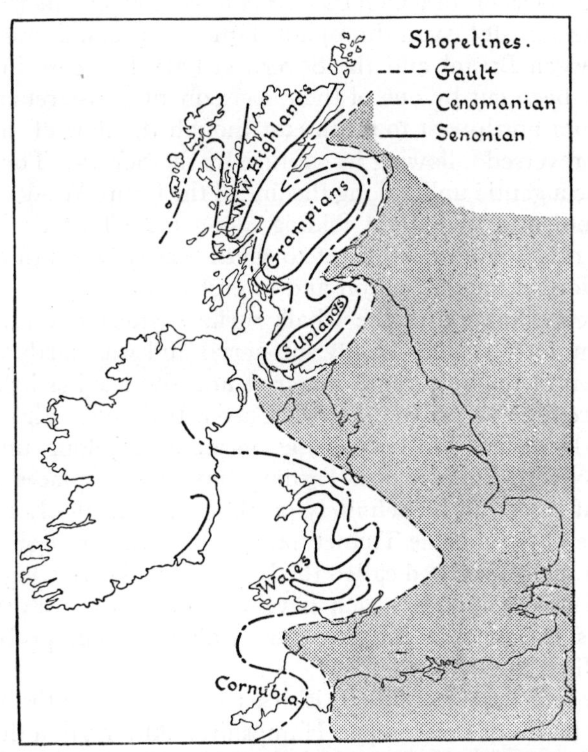

FIG. 54.—The geography of Upper Cretaceous times

sometimes with the other, sometimes separated as the Danian and the Montian. The slight earth-movements which took place were probably the first signs of the great Tertiary or Alpine movements which culminated in Miocene times. Some of the more important post-Cretaceous but pre-Eocene earth movements in Britain include the general uplift in the north and west which not only threw off the Chalk sea but raised much of the chalk surface into dry land and initiated the dominant regional dip of the Cretaceous, Jurassic and Triassic rocks to the south-east. This regional dip resulted both in the development of the English Scarplands and also in the north-west to south-east " consequent " course which can be seen in so many sections of English rivers, sometimes in apparent contradiction to the dictates of the structure of the underlying rocks on to which the rivers have now cut their way. The course of the Bristol Avon through its famous Carboniferous Limestone gorge between Bristol and the Severn estuary is a case in point : it must have been cut by one of these consequent post-Cretaceous rivers flowing from north-west to south-east though the flow of the river has now been reversed following its capture by the Severn. There was also at this time a gentle uplift along the line of the future Wealden anticline or pericline[1] and also slight folding along old Charnian lines—with anticlinal ridges from north-west to south-east lying south of Ipswich, from Watford through London and in east Kent.

The peneplanation of the Chalk surface which naturally resulted from this uplift was thus mainly sub-aerial and was partly effected by the solution of the calcareous material and the leaving behind of the insoluble residue as the well-known Clay-with-Flints. This dissolution of the chalk surface has been going on for a very long time ; much of the Clay-with-Flints is post-Tertiary but in some places, notably in Belgium, it is seen to be definitely pre-Eocene. In the London Basin, at the base of the marine Thanet Sands where they rest on the Chalk, there is an interesting bed called the Bullhead Bed consisting of unworn flints in a clayey matrix which is very similar to the ordinary Clay-with-Flints except that the flints are stained green, probably by a mineral allied to glauconite.

The Eocene deposits of the British Isles are almost entirely restricted to the two basins of London and Hampshire. Although in their present form these two are synclinal basins formed by Alpine folding after the deposition of the Tertiary beds and in which the Eocene strata have been preserved from denudation, the synclinal hollows were already present in embryo when the Eocene sea gently and quietly invaded the old chalk surface. Thus the London Basin and the Hampshire Basin are also basins of deposition, from the denuded margins of which the strata have been removed but round which it is at the same time possible to trace approximately the old shore lines. The geography of

[1] Pericline—an anticline pitching at both ends. See Plate IIIA.

the period may be summarised by saying that a partly enclosed sea, which has been called the Anglo-Franco-Belgian Basin or the Anglo-Gallic Basin, covered the south-east of England, the north-east of France and the greater part of Belgium. The uplift of the Weald must have formed a long low island or at times a submerged shoal partly separating the two main areas of deposition in England. Into this sea there emptied at least two large rivers, bringing masses of sediment.

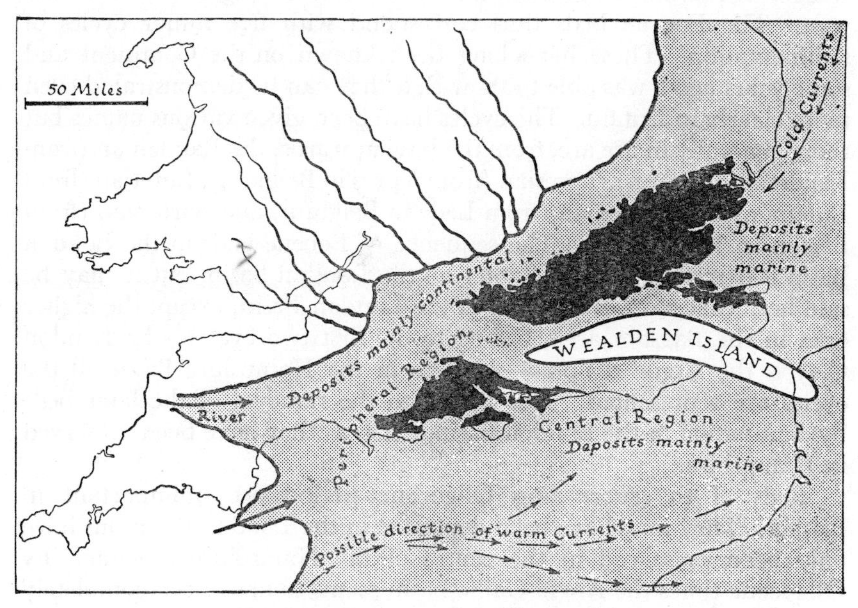

FIG. 55.—The geography of Eocene times

One, which has been called the Eocene Amazon, flowed from the west or west-south-west and deposited its load of sediment in the western parts of the London and Hampshire basins. The other, flowing probably from the south or south-east, poured into the basin in the area to the south-east of Paris.

It is to be remembered that the earth movements which culminated in the great Alpine storm had begun as early as the latter part of the Cretaceous and there were intermittent movements throughout the

Eocene. It appears that the most important of these movements was the periodic increase in the uplift of the Wealden dome which was automatically accompanied by a deepening of the London and Hampshire depressions. Consequently, the history of deposition in each of these basins is their filling up with river-borne sediment from the west with a periodic invasion by the sea from the east. The strata are arranged in wedges—wedges of continental (fluviatile, deltaic or estuarine), strata thick in the west and thinning out to the east, and wedges of marine strata thick in the east and tailing out towards the west. The Eocene beds thus correspond with five minor cycles of sedimentation. These have long been known on the Continent and some years ago I was able to show that they can be demonstrated even more clearly in Britain. The cycles have been given various names but those generally in use are, from the base upwards, the Landenian (from Landen in Belgium), Ypresian (from Ypres in Belgium), Lutetian (from Lutetia or Paris), Ledian (from Lede in Belgium) and Bartonian (from Barton in Hampshire). The sequence of Eocene beds in the London Basin is described in the Chapter on the London Basin, but it may be said here that most of the beds in the London Basin, except the higher ones in the Bagshot area, belong to the first two cycles—the London Clay is the Argile d'Ypres—whereas in the Hampshire Basin all the cycles are represented. This is due to the removal of the later beds by denudation in the London Basin, whereas they have been preserved in Hampshire.

Beds of the succeeding Oligocene period are unimportant in Britain. They may have been deposited in the London Basin but have only been preserved in the centre of the Hampshire syncline. By Oligocene times the Anglo-Gallic basin seems to have been completely cut off from the Southern Ocean or Tethys which occupied the whole area of southern Europe in early Tertiary times.

Both Eocene and Oligocene beds are unimportant in other parts of Britain. There is the small Bovey Tracey basin near Newton Abbott in Devonshire, in which deposits of lignite were formed, and some of the leaf-bearing beds of the Scottish island of Mull, which are interbedded with lava flows, may be Eocene or Oligocene, though most are believed to be Miocene in age.

The Alpine earth-movements are well described, so far as Europe is concerned, as the Alpine storm. With a European centre in the Alps, Britain lay on the fringes of the main area of crustal folding and in

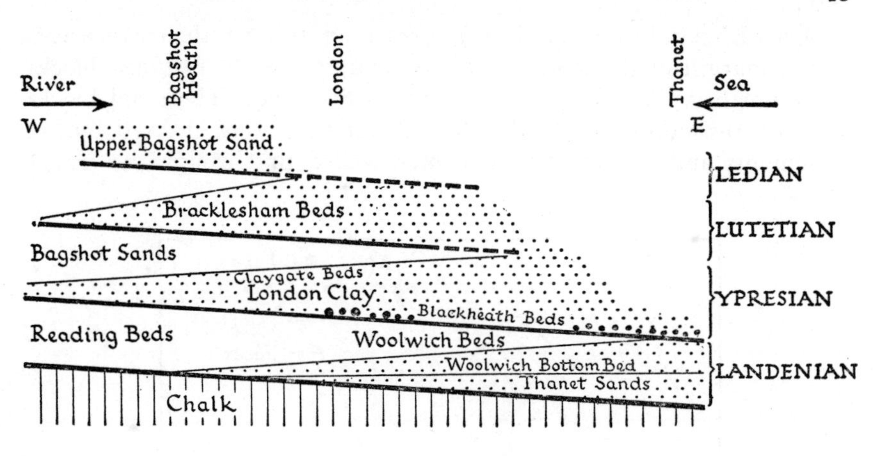

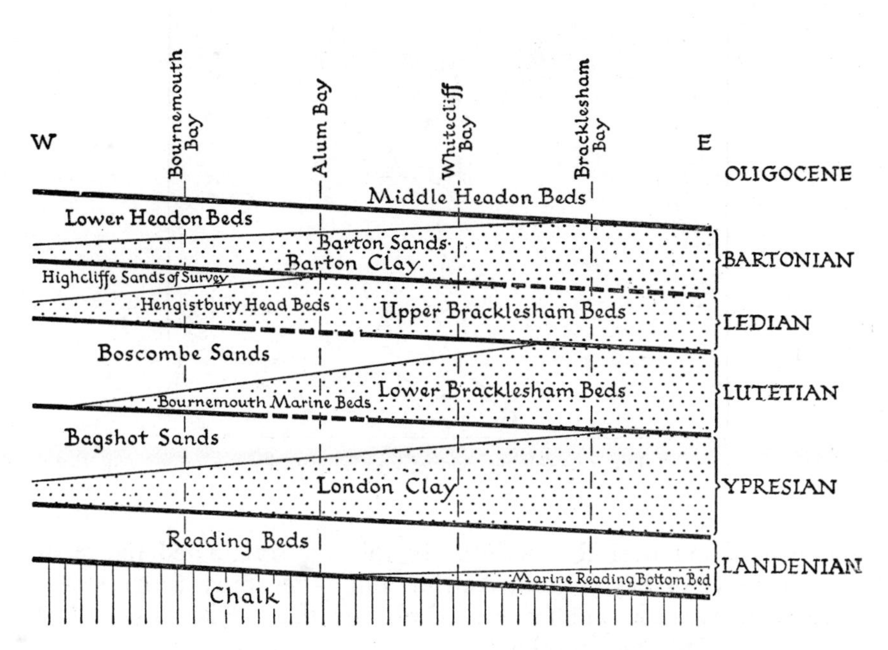

FIG. 56.—Cycles of sedimentation in the English Tertiary
The upper diagram refers to the London Basin, the lower one to the Hampshire Basin.

southern England we see, as it were preserved in stone, the earth-waves on the margin of the storm breaking against the old resistant blocks of the north typified by the Highlands of Scotland. These old blocks were too resistant to be further folded but they were cracked by the movements and masses of molten lava welled up through the cracks

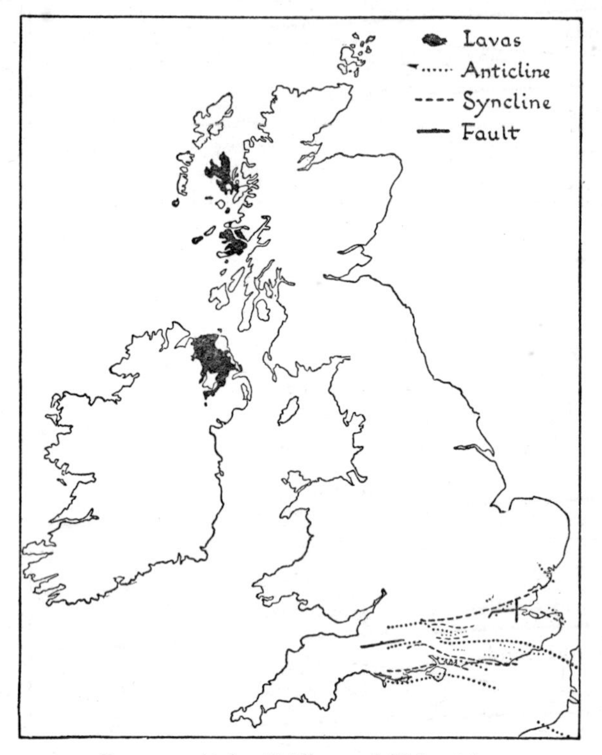

Fig. 57.—Alpine Folding and Vulcanicity

and poured out over the surface. In this way originated the great basaltic flows of Skye, Mull, Antrim and other areas ; the cracks filled with now consolidated lava constitute the " dyke swarms " of northern England and Scotland.

It is of more than passing interest that the area of most intense Alpine folding in Europe coincides with the great oceanic trough of

A Coal Measure forest. From a diorama in the Geological Museum

A Jurassic landscape. From a painting in the Geological Museum

A landscape of London Clay times

The Thames Valley in the Old Stone Age
Both Plates from dioramas in the Geological Museum

deposition in earlier Tertiary times known as the Tethys. The buckling of a geosynclinal trough to form a great land area is to some extent paralleled in miniature in England in the Wealden area. The Weald is bounded on the north by the old Palaeozoic ridge which underlies London and it is probable that a comparable ridge exists somewhere under the English Channel. The Weald is superficially an anticline but underlying it is a much older synclinal basin of deposition. There is thus a superficial anticline superimposed upon a deep-seated syncline in much the same way as the main Alpine folds are superimposed upon a former synclinal basin of deposition. The probable explanation is that the old ridges are relatively stable whereas the basins between have been subject to extensive up and down movement so that at one time they may be deep troughs, at other times lofty land masses.

At the end of Miocene times the structure of Britain as we know it to-day was virtually complete. We have since suffered but minor changes due to earth-folding movements and earthquakes—though there have been many changes of other types due to denudation, isostatic movements and glaciation.

The hills and mountains which were erected in Britain by the Alpine movements were but tiny affairs in comparison with the giant ranges which arose here as a result of the earlier movements—the Caledonian or the Armorican—and do not bear comparison with the huge mountain chains of Alpine date elsewhere which to-day form the major relief features of the earth's surface and were formed at the same time as such minor features as the crest fold of the Isle of Wight. Nevertheless, because it is of comparatively recent date in terms of geological time, the Alpine folding is responsible for many of the well-known features of southern England—such as the monoclinal fold or asymmetric anticline which forms the Isle of Purbeck and the Isle of Wight. It is the almost vertical chalk of the northern limb which forms the ridge, now breached by the sea between the Isle of Purbeck and the Isle of Wight, on which Corfe Castle stands and of which the Needles are a part. The Portsdown Hills at the back of Portsmouth are a minor upfold of chalk ; the Weald itself is mainly of Alpine date ; a tiny upfold gives rise to the knoll of chalk on which Windsor Castle stands. Northwards the folds become less conspicuous but it is in Scotland that Tertiary lavas contemporaneous with the Alpine folding play such an important role. The Isle of Staffa with Fingal's Cave (see Plate XVA) and the celebrated Giant's Causeway of Northern

Ireland are of Tertiary basalts whilst Ireland owes the Mourne Mountains as Scotland does the Cuillins or Coolins of Skye and probably the magnificent Goat Fell of Arran to intrusive granites and other plutonic rocks of Alpine age.

The story since the Alpine mountain-building period is so largely one of denudation that its unravelling is a study for the geomorphologist quite as much as for the geologist and must now be considered in some detail.

Since the publication of the first edition of this book Professor L. J. Wills has published his magnificent *Palaeogeographical Atlas* (1951) to which reference should be made.

It will be noticed that many of the maps given on the preceding pages (Figs. 45 to 55) suggest that the North Sea, in part at least, is a very old and persistent feature with the result that certain beds on the eastern side of England are closely comparable with those on the continent, in Belgium, Holland and Germany. The discovery of very rich gas-fields on the continental North Sea margins brought into prominence the importance of this connection and led to an intensive exploration of the bed of the North Sea by drilling from platforms erected on the sea floor. Geological theory was justified and from 1965 onwards important resources of natural gas were discovered including, in 1966, a strike in north Yorkshire some eight miles inland from Scarborough. The Groningen gas of Holland comes from Lower Permian Sandstones (see Fig. 46) which probably extends under most of the North Sea.

THE PLIOCENE PERIOD

FOR LONG it was believed by geologists that Pliocene deposits were to be found in Britain only in East Anglia and in one or two small patches in the south-western peninsula. Although some other beds—mainly high-level gravels—have been recognised since as Pliocene or late Miocene in age it is true that the interest in the geology of the period is limited so long as attention is restricted to the deposits which have been preserved. It was, however, during this period that the forces of denudation, both sub-aerial and sub-marine, were at work carving the surface into the rough semblance of what it is at the present day. True, much of the detail was destined to be modified by the ice sheets of the Great Ice Age which followed, but the application of the intensive study of land-forms and their development has revealed how much the present scenery owes to features which were sculptured in Pliocene times. In contrast to earlier periods, the geological story is written not so much in the strata which were accumulated as in the sequence of events of denudation.

As we have seen, the main Alpine storm culminated in the Miocene and, although in other parts of the world folding movements continued along the same axes well into the Pliocene, it is probable that in Britain folding had virtually ceased by the beginning of the Pliocene. If so, subsequent movement must have been of a eustatic nature—an up and down movement of the land relative to sea level—accompanied only by slight " warping " along the lines of the earlier folds.

The broad picture which we should visualise is of southern or south-eastern England ridged up into a series of folds, wave-like in form and tending as it were to " break " towards the north. The folds are most marked and sharpest in the extreme south, becoming more gentle northwards. The sharp southerly ones are typified by the central fold of the Isle of Wight and Isle of Purbeck. By Pliocene times their crests were being denuded away and, what is of the utmost importance, the level of the sea then stood in what is now southern England at several hundred feet above its present level. Much of the denudation was thus sub-marine, and one of the great legacies we have from the

Pliocene is of wave-cut platforms or sub-marine peneplanes (see Plates 4, 5A and XXB) standing at various heights above sea level. These peneplanes or platforms at different levels mark stages in the intermittent uplift of the land relative to the sea. Northwards—especially north-westwards—the level of the surface was higher and it is probable that most of Scotland was dry land. The drainage of the land masses was naturally down the predominant slope and we can picture rivers draining from north-west to south-east and from north to south, and so along many of the river valleys with those trends which are conspicuous in Wales and other areas at the present day. It will be remembered that many of the Eocene rivers flowed in these directions and it is therefore probable that the Pliocene rivers followed and deepened pre-existing valleys (see Fig. 69).

At the beginning of the period south-eastern England was between 650 and 700 feet *lower* than at present, and the crest of the Wealden dome must have formed a low island from the surface of which the chalk had already been removed to expose the underlying rocks. The Pliocene sea occupied the synclinal basin or downfold of what is now the London basin to the north. Deposits from this early Pliocene sea are preserved as patches of reddish sands and gravels high up on the North Downs—especially above Lenham in Kent and hence called the Lenham Beds. They are found largely in solution pockets or " pipes " in the chalk surface at about 650 feet above sea level and have yielded sufficient fossils to prove their age. Some palaeontologists wish to claim the beds, on the basis of the assemblage of fossils, as late Miocene rather than early Pliocene but this makes little difference to the general picture. Similar beds, in which fossils have not been found but which from their position are almost certainly of the same age, occur at intervals along the North Downs, notably at Headley Heath and Ranmore Common (Surrey). There they are coarse gravels with little-rolled flints and sharp sand and, like the other deposits, suggest deposition in shallow water near a shore-line. Following up this clue, Professor S. W. Wooldridge succeeded in tracing the actual shore-line in a number of places. It must have been marked originally by a line of low cliffs, and although the sharp cliff line has been smoothed by later denudation there is still a sudden break of slope high up on the North Downs to mark its position. The line of cliffs is not continuous ; in some places it lay *south* of the present crest of the North Downs and has been completely denuded away. These Lenham Beds are found in

L. D. STAMP

The Malvern Hills from the north-east
An Armorican mountain range rising from a Triassic plain (See page 123 and page 220)

The Pass of Llanberis, North Wales L. D. STAMP

A typical U-shaped glaciated valley, interrupted by a rocky spur due to an outcrop of hard dolerite (See page 87 and page 221)

comparable positions on the other side of the Strait of Dover and are there known as the Diestian—they form the capping of low hills known as Les Noires Mottes, on the top of the chalk near Calais. An interesting method of analysis has been applied to these various patches of sand in order to prove or disprove their contemporaneity. Whilst the majority of the grains in a sand consist of quartz there are scattered grains of other minerals. These are usually heavier than the quartz grains and can be separated out by a process similar to the process of panning for gold ; the concentrates of heavy mineral grains can then be examined under the microscope and the minerals identified. Beds of the same age tend to have a characteristic assemblage of heavy minerals, differing from those of other horizons and the consequence naturally of the derivation of the sands from a particular source or direction. For example, in the Lenhamian-Diestian Beds grains of the rare mineral monazite are always present whereas this is rarely found in sands of other ages with which the Lenham beds might be confused. Limited as they are in extent the Lenham Beds have a special interest to the naturalist of to-day. They give rise, as in the case of Headley Heath, to level areas of sandy, podsolic, acid soils high up on the chalk downs : hence the remarkable contrast between the heather-bracken-birch of Headley Heath and the rolling fescue grasslands of Epsom Downs only a mile or two away. Even when all traces of Pliocene deposits have been swept away, where the chalk surface was subjected to sub-marine peneplanation by the Pliocene sea, it presents smooth outlines. This is well seen in East Kent.

Whether exactly contemporary with the Diestian Sea or not it is impossible to say, but about this time the sea covered most of Devon and Cornwall. The higher parts of Dartmoor, Bodmin Moor and Exmoor must have stood out as islands, with the various " tors " as imposing cliffs like those of present-day Land's End, whilst all around the waves were cutting the familiar smooth platform typical of submarine peneplanation. The land rose at intervals and hence there is actually a succession of platforms separated by sharp breaks of slope which are the old cliff or shore lines. One of the most extensive and best developed is the 400-foot platform, so-called because it is at approximately that height above sea level. Once again the work of Pliocene seas is of outstanding significance at the present day—the rolling rather featureless plateau which makes up so much of the interior of Cornwall and which many people label " dull " is the result

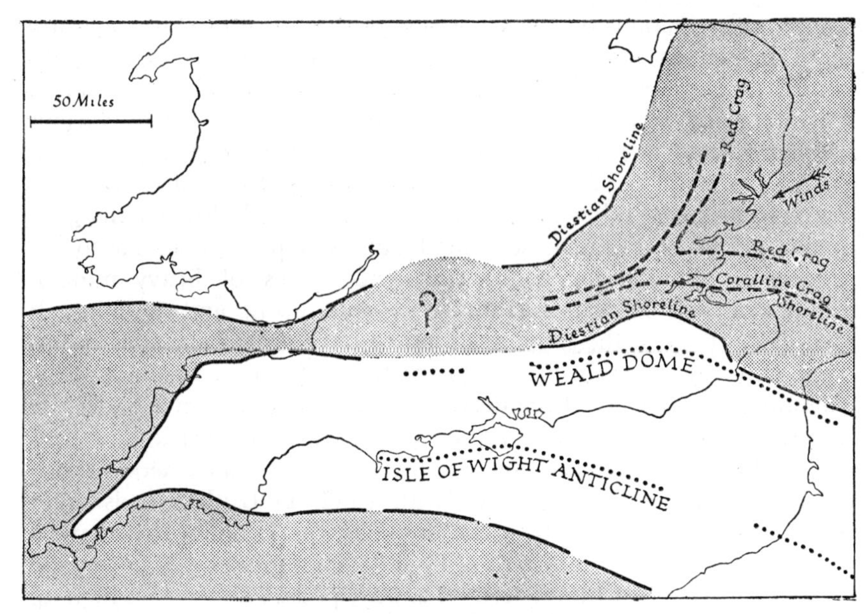

FIG. 58.—The geography of early Pliocene times

of Pliocene peneplanation. It is naturally windswept, hence the lanes between high earth and stone-faced banks which are such a feature of Devon and Cornwall ; hence also the siting of settlements in secluded hollows ; hence also the deep, steep-sided valleys still cutting rapidly into the plateau—indeed all the outstanding scenic features of the south-western peninsula are directly traceable to Pliocene geography. Even the magnificent modern sea-cliffs are where the peneplanes of the past meet the seas of to-day. These features are illustrated in Plates 4, XXVIII and 5A.

Returning to south-eastern England, the next stage was a general uplift of the land and a slight warping which deepened the London basin. Conditions remained stationary long enough for the waves to cut a marked platform which lies at about 400 feet above present sea level and on which are found gravels of well-rolled pebbles. These are the well-known plateau gravels of the Chilterns whose origin has been long and hotly disputed : other patches occur at the same height, for example those capping Shooter's Hill, Hampstead and Highgate, and play a leading part in the siting of many of the favourite open spaces of greater London. The fact that these gravels are found at such diverse places always at about the same level shows that warping of the surface had ceased and that later movements have been eustatic —simply up and down. At the time of formation of the 400-foot plateau gravels the open Pliocene sea lay to the north-east and deposits of Pliocene age are found over a considerable area in east Suffolk and the adjoining parts of Norfolk. These " crag " beds are sometimes of the nature of shell-banks accumulated a short distance offshore. As the land rose and the sea retreated to the north-east the later deposits are not superimposed on the older but are found further away from the London area. Apart from rolled sandstone pebbles with Miocene fossils (the Boxstones, so-called because many when split open reveal a fossil inside) the oldest Crag is the Coralline Crag, made up of immense numbers of fragments of what are actually polyzoa or sea-mats and their allies—formerly called corallines. In some parts of the Crag beds there are immense numbers of molluscan shells. These are of warm-water species showing that the way was open at the time through central Europe to warmer waters. The Red Crag follows and in it shells of northern origin become steadily more and more important at the expense of southern ones ; of the molluscan fossils of the higher parts no less than 89 per cent are of living species. The Norwich Crag, which rests directly on the Chalk further north, is believed to be part of the deltaic deposits of the ancient Rhine which emptied into the North Sea somewhere in the neighbourhood of Cromer. The Chillesford Beds which follow have a restricted and curious disposition : it is claimed that they mark the actual course of the lower Rhine. The youngest Pliocene beds of East Anglia are the Cromer Forest Bed Series of two freshwater horizons of peat with land and freshwater shells separated by an estuarine clay with many remains of land animals. Unlike the molluscs, these are of extinct species—the mammoth (*Elephas primigenius*), *Elephas antiquus*, woolly rhinoceros, musk-ox, sabre-toothed tiger, and cave bear. This is a curious mixture of southern and northern forms and may indicate the oncoming of cold conditions, but there is a possibility that some of the bones may be " derived " or washed out of an earlier deposit.

Great excitement in scientific circles was caused some years ago by the discovery at the *base* of the Norwich Crag, by the late Mr. Reid Moir, of what he claimed to be implements fashioned by man. If they

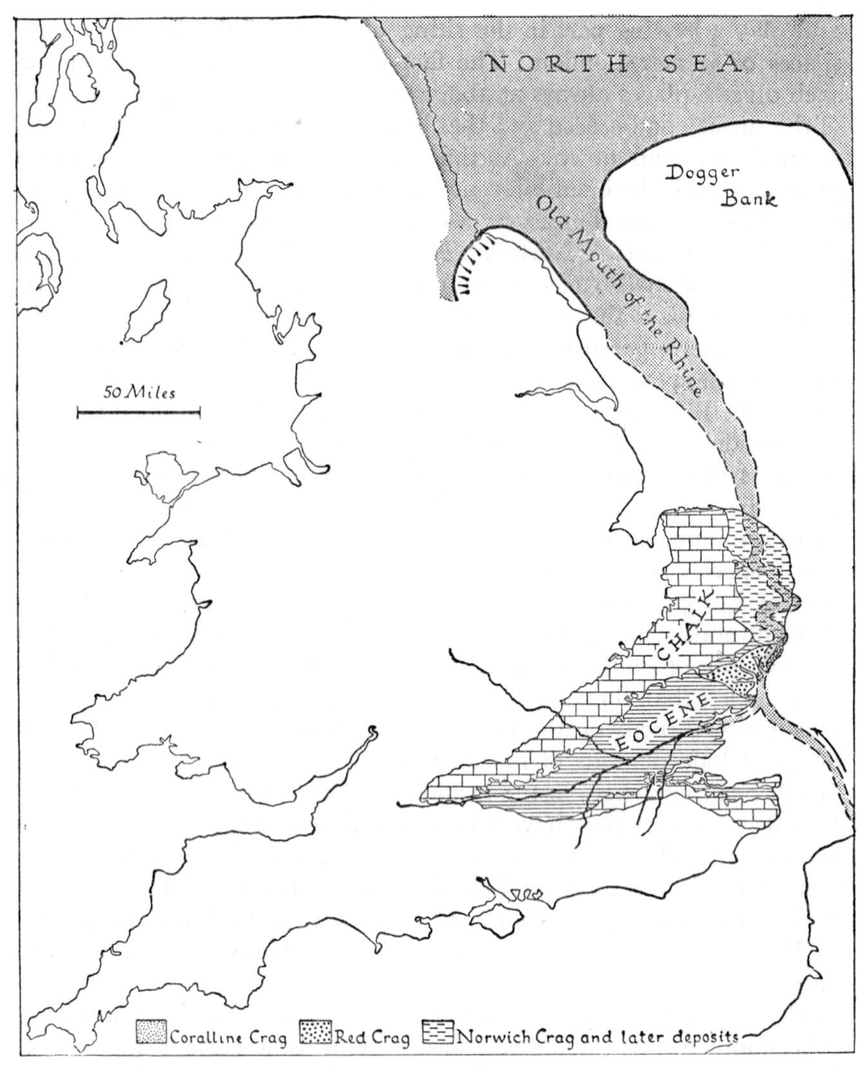

FIG. 59.—The geography of the late Pliocene showing the course of Chillesford River (probably the Rhine)

are in truth artefacts—and expert scientific opinion is still divided—then there were beings in Britain sufficiently human to be fashioning crude implements in late Pliocene times.

Returning to the London basin proper, there is a further platform now traced over a wide area at about 200 feet above sea-level and which is probably late-Pliocene in date. It may be contemporary with the Red Crag, in which case the pre-Glacial river valleys cut into it would be of tributaries of the old Rhine.

What can we say of the geography of other parts of Britain during the Pliocene period? The answer is, remarkably little, though there is probably a great deal waiting to be discovered by an intensive study of the land forms. To take a single example, we have seen that the main peneplanation of the mountains of North Wales probably took place in early Cretaceous (pre-Chalk) times. To-day that peneplane can be detected only with difficulty through the concordant summit-levels of the mountains. The chalk and other deposits, if such were ever deposited, have been completely removed and an enormous amount of denudation has taken place. The tens of millions of years of the whole Tertiary era have not, in fact, been too long for the process and one may presume that the work of denudation was going on continuously during the Pliocene and some day it should be possible to trace it stage by stage.

Since this chapter was written for publication in the first edition (1946) much research has been carried out on the Pliocene and Pleistocene deposits of eastern Britain, and by modern methods it is possible to date the various beds much more exactly. It is now generally agreed that the Newbournian Red Crag, the Norwich Crag, Chillesford Clay and Cromer Forest Bed should all be regarded as Lower Pleistocene. See also p.171.

THE GREAT ICE AGE AND AFTER

ALTHOUGH the Ice Age is the last great episode in the geographical evolution of this country and in terms of geological time took place only recently, it has proved extraordinarily difficult to reconstruct the exact sequence of events during and since the time when ice-sheets overwhelmed the whole of the British Isles except the south. The evidence is fragmentary and not infrequently appears to be contradictory—the deposits left by the retreating ice were mainly easily eroded sands and gravels and even the boulder clay, not so easily removed by ordinary denudation, left by one glacier or formed during one cold period may have been swept away by the next. Further, the records left in one part of the country by one ice-sheet or glacier may be extremely different from those left by another but nevertheless contemporary ice-sheet in a different part. Then there is the extreme difficulty of correlating deposits in the non-glaciated parts of the country—and it is naturally in these that fossils, both animals and plants, are to be found, as well as implements fashioned by early man. Many of the facts established by patient field and laboratory investigation are capable of more than one interpretation. For example, the conversion of a large body of water into ice may cause a general lowering of the water level in an enclosed basin or sea and so produce the same effect as a rise of the land surface. In recent years many advances in knowledge have been made by the application of methods—such as the microscopic analysis of remains of pollen in successive layers of peat—not previously used. Far more has become known even in the last few years through the methods of geomorphological analysis—the detailed study of land forms. A raised beach, a tiny fall in an otherwise slow-moving river, wisps of gravel at different levels—all these have been made to yield a continuous story not previously believed possible.

It must, however, be made clear that there is, as yet, by no means complete agreement on the sequence of events. Yet from one point of view the details are of the utmost importance and interest to the naturalist and that is in regard to the development of our native flora and fauna. The oncoming of the extreme cold of the ice age drove out the existing animals and plants, leaving but a few to survive in the unglaciated regions of the south. On each retreat of the ice there was a repopulating of the islands by successive waves of migration from the Continent. As soon as Britain was finally cut off by sea from the Continent this migration either ceased or baceme increasingly difficult so that the date of the final opening of the Strait of Dover is a matter of the greatest interest. One must say " final opening " because there is evidence of earth movements which may have alternately cut and joined the land masses. Similarly the date of the separation of such outlying groups as the Shetlands, Orkneys, and Outer Hebrides as well as of Ireland itself is of first-class importance, yet in many such cases the geological evidence is extremely scanty.

Geologists of the earlier part of the nineteenth century were faced with the problem of proving that the " drift " deposits scattered about the surface of Britain were indeed the products of ice-action and not of Noah's Flood—the name " diluvium " still occasionally applied to them is a reminder of the old controversy. This was followed later by the long arguments between the monoglacialists who considered there had been only one ice age, and those who found evidence for a succession of ice ages separated by relatively warm interglacial periods. The great geologist Sir Archibald Geikie was an exponent of the former school ; his brother Professor James Geikie, maintained the existence of evidence of seven glaciations. The arguments which James Geikie adduced for the existence of more than one ice age in the 3rd Edition (1894) of *The Great Ice Age* were amply vindicated in the great work of A. Penck and E. Brückner on the Alps (*Die Alpen im Eiszeitalter*, 1901-1909) which established without doubt a fourfold advance and retreat of the Alpine glaciers. The four glaciations were called the Günz, Mindel, Riss and Würm glaciations, and each was followed by a milder or interglacial period. For long this work coloured all investigations in Britain and geologists here did their utmost to fit the facts into the continental chronological scale. Actually British conditions are more closely linked with those in northern Europe—on the German plain and in Scandinavia—and, in turn, work there has rather overshadowed investigations at home. For reasons which will be made clear later in this chapter, there is the utmost danger in attempting to fit the observed facts in Britain into any chronological scheme based primarily on continental conditions, though broadly the evidence in

Britain tends to the acceptance of four main glaciations separated by three interglacial periods of which the middle one seems to have been much the longest.

To appreciate the conditions under which the Pleistocene or Glacial Period opened, we may recall that at the close of the Pliocene the land surface in the south-east of England, although it had been steadily rising, was between 150 and 200 feet below its present level and that the late Pliocene rivers were cutting channels in the platform which now lies about 200 feet above sea level. The Thames was already following something of its present line, but curving away north-eastwards to join the Rhine, the united rivers emptying into a shallow North Sea somewhere in the neighbourhood of present-day Cromer. The Cromer Forest Bed series which we mentioned as the youngest Pliocene beds in Britain are claimed by some as early Pleistocene, but the changes in the geography as well as in the flora and fauna about this time were so gradual that it is difficult to know where the line should be drawn between the two periods.

Though the detailed investigation of glacial deposits in all parts of the country is far from complete and the need for more and ever more careful and detailed observations offers a field of study for the amateur geologist—it is not too much to say with Professor L. J. Wills that we owe the greater part of our knowledge of the glacial history of Britain to the amateur rather than the professional geologist—numerous attempts have been made in recent years to attempt to summarise the present state of knowledge. We may refer in particular to the Second Edition (1937) of the late W. B. Wright's *The Quaternary Ice Age* and to the Presidential Address to Section C of the British Association at York in 1932 when Professor P. G. H. Boswell attempted what he described as " a word-picture of Pleistocene conditions in the British area in so far as they affected the occupancy of the country by Early Man." Boswell naturally based his survey on his own very wide knowledge of conditions in East Anglia ; Professor L. J. Wills in his turn in 1937 summarised the evidence provided by the west Midlands. In the meantime Professor W. B. R. King and Dr. K. P. Oakley had applied modern methods of study to a reinterpretation of the evidence afforded by the terraces of the lower Thames valley (*Proceedings of the Prehistoric Society*, N.S., 2, 1936, 52). More recently Dr. W. J. Arkell, from the starting point afforded by a study of the raised beaches of southern England (particularly the " Head " (see p. 89) of the coasts of the

L. D. STAMP

Mosedale in September (near Wasdale, Lake District) (See page 87 and page 222)

PLATE 20

L. D. STAMP

The marshes of Afon Glaslyn, looking north from Portmadoc (See page 48)

south-western peninsula) has assayed a new correlation of Pleistocene deposits. When he read his paper before the Geologists' Association of April 2nd, 1943, he used the continental terms—Günz, Mindel, Riss, Würm—for his episodes but before the paper came to be published (*Proceedings of the Geologists' Association*, 54, 1943, 141-170) he had found it necessary to introduce a completely new terminology. This is quoted not in criticism but to indicate the uncertainty which still exists regarding the detailed history of Britain during the Ice Age—a period of the utmost importance in the immigration and establishment of our fauna and flora. With regard to the post-glacial history, the pioneer work of Dr. H. Godwin on pollen analysis in successive peat deposits has put this study on a firmer basis but there is still need for clearer thinking on the geographical aspects. In a survey of the present state of knowledge which I attempted to make a few years ago on the geographical evolution of the North Sea basin, it was one of my objects to stress the importance of the opening of the Strait of Dover not only because of the direct effect of cutting the land bridge between Britain and the Continent but because of its indirect effect in permitting the free circulation of ocean water round the British Isles. It is my belief that the abrupt change from a cold or "Boreal" climate to the milder, damper "Atlantic" climate was occasioned by this fundamental change in oceanic circulation and its consequent effect on air movements.

Having thus attempted to emphasise that any reconstruction of conditions during the Pleistocene must still be very tentative, we may quote from Boswell's word-picture. At the end of Pliocene times he pictures a land-area populated by plants living under temperate conditions very like those of the present day and drained by a great river flowing northwards to the proto-North Sea and carrying with it the remains of such southern warmth-loving animals as the straight-tusked elephant (*Elephas antiquus*), hippopotamus and rhinoceros (*R. merckii*). The sea into which it discharged, however, was open only to cold northern waters and was inhabited by a molluscan and other fauna essentially typical of arctic or sub-arctic waters.

Then, as now, the highest land of northern Europe was in Scandinavia and it was there naturally that the ice began to accumulate and the first great ice-cap began to form. It grew to a great size and thickness and its accumulation seems to have been accompanied by a marked depression of the land and its consequent submergence. How far this was due to actual weight of ice is still disputed. The depression of the land surface seems to have been general over northern and north-western Europe at least and so to have affected the whole British Isles. The well-known and clearly marked wave-cut platform or raised beach some ten feet above present high-water level so clearly seen round the coasts of Devon and Cornwall is commonly referred to as the Pre-Glacial Raised Beach and is believed to belong to this early period of submergence, though it is only right to mention the doubts recently expressed by Arkell. The Wash and the whole of Fenland was probably occupied by a great arm of the sea, whilst some ancient sea-cliffs such as that at Sewerby near Bridlington belong to this immediately Pre-Glacial time—or rather to the time when the first great ice-cap was forming in Scandinavia. In due course the ice stretched out across the North Sea, partly floating, partly scraping along the bottom as the Great Antarctic Barrier does at the present day. It impinged on the coasts of Durham and Yorkshire, depositing what has there been called the " Basement Clay," and rode over the low ground of eastern Norfolk, there depositing the North Sea Drift. Quite naturally icebergs with boulders on their surface or frozen into their mass would break off from the main ice-sheet and it is these which carried Scandinavian rocks to the shores of the Moray Firth, the Orkneys and further afield. It is possible that the Wash-Fenland depression was also deep enough for ice to float in and drop detritus. Man was living in Britain at the time of the oncoming cold but the implements he produced were of the very primitive types loosely grouped as "pre-Chellean." Until it was proved that " Piltdown Man " (*Eoanthropus*) was a hoax, it was believed that he was the maker of these early flint tools. With the oncoming cold the higher hills must have been covered with snow and it is probable that some of the so-called plateau gravels and irregular drift deposits of the Chiltern dip-slope were produced at this time by solifluction—the movement of a glacial sludge. Wills inclines to the view that the first development of an ice-sheet in Wales—the First Welsh Glacier that spread on to the Midlands—was contemporary with the North Sea glaciation.

The retreat of the Scandinavian ice appears to have been followed by a long period of relative warmth—the First Interglacial. During the earlier part of this interval East Anglia was occupied by a shallow sea or gulf, inhabited by a cold molluscan fauna, into which were discharged large quantities of sand and gravel released by the melting ice-sheet. A general uplift followed and in the Midlands and the upper Thames valley the rivers began to cut down through the high level outwash gravels and to form the highest terraces—such as those between 200 and 300 feet in the Oxford area and along the middle and lower Thames. Presumably man—probably the maker of implements of Chellean type of which derived and abraded specimens are found in later deposits—advanced into and occupied those parts of Britain which were available. Plate 16B is a representation of conditions as they are believed to have been in the Thames Valley at this time—the straight tusked *Elephas antiquus* can be seen on the right.

The First Interglacial interval was brought to an end by the development of " home-grown " ice-caps on the mountains of Scotland, the Lake District, the Pennines and the Welsh mountains. As at the present day the moisture-laden winds doubtless came across the Atlantic Ocean and the main mass of snow would naturally fall just to the *east* of the mountains. Further, as at the present day, the west would be the milder side, the east the colder side, facing the pre-existing Scandinavian ice. Consequently the chief glaciers produced in Britain were on the eastern side of the country where they flowed eastwards until, it would seem, they met the ice which still blocked the North Sea and were forced to turn south or north—parallel to the coasts. Amongst the characteristic rocks transported by the glaciers which enable their courses to be traced may be mentioned the Shap granite of the Lake district—taken as boulders across the Pennines to the eastern side—and the Cheviot granites, transported far to the south. It seems probable that the main stream of ice was split into two by the obstacle offered by the Cleveland Hills of north Yorkshire, and further south there were two main streams of ice on either side of the Lincolnshire wolds, flowing to the south and south-east roughly parallel to them. The western glacier came down the Vale of York across the broad vale of the lower Trent and then fanned out over the Midlands as far south as Moreton-in-Marsh, coming there into contact with ice of Welsh origin (the Second Welsh ice-sheet). It also swept south-eastward, pushing masses of chalk and Jurassic material across the low ground of the Fens and into East Anglia as far south as the outskirts of London and mid-Essex. In East Anglia it came again into contact with the ice which had travelled along the eastern side of the Wolds, across the Wash and so to Norfolk and Suffolk. It has been well called the Great Eastern Glacier. Over the greater part of England this second glacial episode was thus

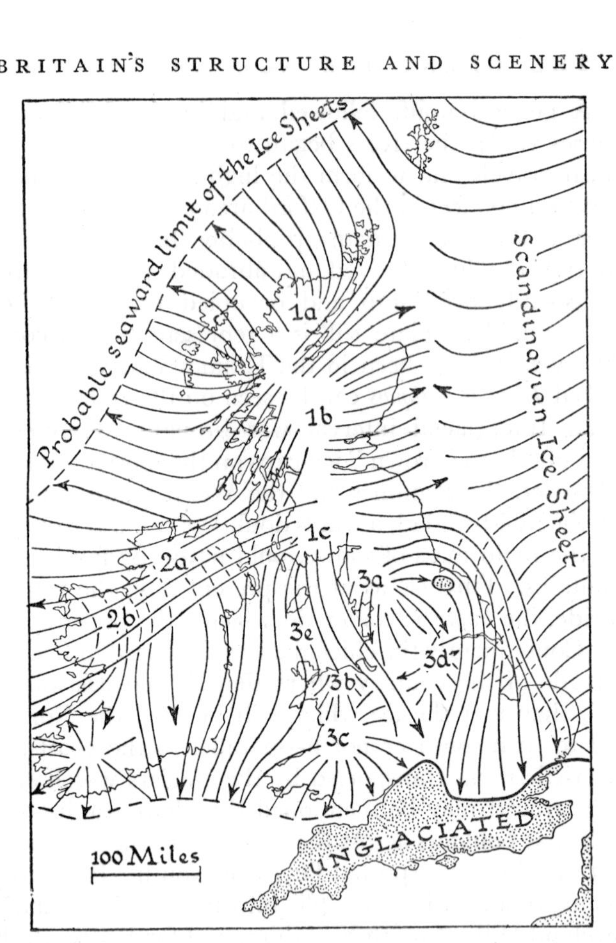

FIG. 60.—The geography of the period of the maximum glaciation in the British Isles

This map shows the approximate position of ice sheets at the second glaciation. The home ice-caps—centres of ice accumulation and dispersal—are numbered. Lines and arrows show direction of ice flow. Where lines are broken the earlier directions were superseded by those shown by solid lines.

the period of *maximum glaciation* during which the great stretch of the Chalky Boulder Clay was deposited.

To go back in time for a moment, when the Scandinavian ice-sheet stretched right across the North Sea what happened to the waters of the northward-flowing Rhine and its tributary the Thames? It is difficult to avoid the conclusion that their waters were ponded back till they reached such a level that they cut an overflow channel south-

Plate XVII

THE ALETSCH GLACIER, SWITZERLAND E.N.A.

This is the type of valley glacier which formerly occupied the valleys of Scotland and Wales. On the right is a glacial lake held up by a moraine stretching across an overflow channel

Plate XVIII

SOIL EROSION IN ENGLAND

J. V. SPALDING

Gully erosion of a cultivated field after a heavy rainstorm on April 28th, 1940, at Nettleton Hill, near Caistor, Lincolnshire

wards thus making the first breach between the Continent and Britain—in other words the first cutting of the land bridge along the line of the Strait of Dover. Quite possibly the elevation of the land which marked the First Interglacial, combined with the partial if not complete disappearance of the ice from the North Sea, resulted in a restoration of the land bridge and the reopening of the normal northward course of the discharge of the Thames and Rhine. Some evidence to this effect is afforded by a study of the deposits on the floor of the North Sea.

Whatever may be the exact interpretation of the various deposits in detail, it is generally agreed that our British glacial deposits can be divided into the " Older " and the " Newer " Drifts. The *Newer Drifts* may be recognised by the freshness of their surface features which exhibit clearly moraine-ridges, eskers, kames, kettle moraine, and meres or glacial lakes in boulder-clay hollows. It is possible to trace a close relationship with existing drainage, to delineate the shore-lines of former lakes and to detect the overflow channels by which their waters drained away. The Newer Drifts are less extensive than the older and so the areas of *Older Drift* outside the tracts covered by the Newer have been subjected to extensive denudation both under normal sub-aerial conditions and under the conditions of intense cold which must have persisted round the margins of later ice-sheets. The Older Drifts are the legacy of the first two glaciations we have just considered. When the Great Eastern Glacier retreated with the development of the warmer conditions of the second inter-glacial much of the Midlands and east of England was destined never again to be actually covered by ice. The rivers there, swollen by melt-waters, deepened and enlarged their valleys and the well-known 100-foot or Iver-Boyn and Swanscombe terraces of the lower Thames belong to this period. The country was inhabited by the makers of Acheulian types of flint implements. River-Drift man, as he is sometimes called, wandered over the country camping near rivers and meres, where his flint-flaking sites may still be found, and doubtless engaged in the hunting of the southern mammals—the straight-tusked elephant, rhinoceros and hippopotamus—which again roamed the land. In the lake deposits are found a rich fauna and flora including species (such as the little mollusc, *Corbicula fluminalis*, now only found in the Nile and certain other warm rivers) typical of conditions warmer than those now found in Britain. But Acheulian implements are not found in the north of England and one may infer that the conditions there remained too cold. There is evidence, too, that strong winds prevailed round the margins of the retreating ice-sheets and wind-facetted pebbles are found in some of the gravels.

This second interglacial period is believed to be the longest. Towards the latter part of it the British area seems to have been invaded by a different type of man—a short, broad fellow with very pronounced brow-ridges over his eyes—who made implements of a kind quite different from the Acheulian. Neanderthal Man, as he is called (*Homo neanderthalensis*) was the maker of the Mousterian implements which are finely chipped on one side and along one edge only. They were made from large *flakes*, not from flint *cores* like the earlier ones. Mousterian man seemed to have supplanted Acheulian man (who was incidentally, as far as the little evidence we have goes, of a more modern type in appearance) but very soon, with the oncoming once more of colder conditions, was driven to live in caves and hence is often called Cave Man. He was forced to dispute the occupance of the few available caves with the cave-bear, cave-lion and cave-hyena which, despite their names, were denizens of open lands further south and like man only retreated to caves to escape the cold.

The Third Glaciation was marked by a southward re-advance of the ice—actually in three main lobes. The largest was that which again followed the east coast and as the "Little Eastern Glacier" impinged on the north Norfolk coast where its terminal moraine forms the Cromer Ridge. The second ice-sheet was that which flowed down the Vale of York. The third was the very interesting ice-sheet which seems to have originated from the Southern Uplands of Scotland and spread across the shallow Irish Sea—pushing before it or carrying frozen into its mass sand and pebbles with sea-shells from the floor of the sea—and then overrode the low coast of Lancashire and Cheshire, eventually reaching south right into the Midlands, there to cause much interference with the pre-existing courses of rivers.

Doubtless it was through the agency of icebergs broken off from the Irish Sea ice where it terminated between the coast of Wales and Ireland that boulders were carried and dropped round the Scilly Isles where they are preserved in some of the raised beach deposits.

In the north of England cold conditions were more continuous than in the south and it is often not easy to separate the boulder clay left by the Third Glaciation from that left by the Second. Thus in Yorkshire it is sometimes possible to distinguish an Upper and a Lower Purple Boulder Clay but only if the two are separated by intervening sands and gravels. Even further south the Upper Chalky Drift is not by any means easy to separate from the Lower Chalky Boulder Clay.

In the south the Taplow or 50-foot Terrace of the Lower Thames is assigned to the latter part of the Second Interglacial and the gravels of this terrace are covered with the well-known Brickearth. This is very like the well-known loess of the continent of Europe which is a wind-borne deposit. It was probably deposited by the winds which swept over the country on the approach of the Third Glacial period—and later. In the moister conditions of Britain it is possible that the loess dust was at least partly deposited in water. In the lower beds of the Brickearth are found both the implements of Mousterian man and the remains of that typical denizen of the cold regions, *Elephas primigenius* (the mammoth). The Third Glacial was marked in the south by the formation of much glacial hill sludge, so that the lower deposits of Coombe Rock (with Mousterian implements) belong to this period, which was accompanied by a sinking of the land.

Curiously enough the Third Interglacial is more difficult to trace in its effect but, as Boswell has said in his summary, " By inference, the ice must have retreated on a large scale, for Aurignacian man was able to establish himself on many sites and to reach the caves of Derbyshire and North Wales and to leave, in the former case, examples of his *art mobilier*. He was accompanied by a fauna of arctic and tundra type. Aurignacian man was definitely of modern type (*Homo sapiens*). His beautifully chipped stone knives show a very high level of skill and his art was of a vigorous and striking character. The situation of his " floors," at present below sea level at some localities, indicates that the land area stood higher than now so that communication with France must have been relatively easy." Accordingly, the earlier Coombe Rock in the south was subject to erosion, the " deep channel " below the present base level of the Thames was excavated and there are similar buried channels associated with the upper Thames (Oxford), Severn and Worcester Avon. It is to be noted that in his recent summary Arkell attributes the Aurignacian buried channel period to an " interstadial " period in the midst of the Fourth or Final Glaciation rather than to the Third Interglacial. Though there is not, therefore, full agreement, it is generally accepted that after the period

CHRONOLOGY OF THE GREAT ICE AGE AND AFTER

	Climatic and Ice Conditions	*Human remains and cultures*	*Important events and deposits in Britain*
Post-Glacial	Sub-Atlantic Atlantic (wet, mild) [sudden change] Boreal (dry: cold winters, warm summers) Pre-Boreal (dry, cold) Sub-Arctic	Historic Times Iron Age Bronze Age Neolithic (New Stone Age: polished implements)	VIII Alder-Oak-Elm Birch (Beech) Zone of Godwin VII Alder-Oak-Elm-Lime Zone Final cutting of the Strait of Dover VI Pine-Hazel Zone V Pine Zone Steppe Animals IV Birch-Pine Zone (sub-merged forests) Tundra conditions Many glacial lakes
Fourth Glacial	Lake District and other mountain regions with valley glaciers	Magdalenian (Creswell Cave Man) Aurignacian	York Moraine Last Coombe rock formed in south Ponder's End Arctic Plant Bed
Third Inter-Glacial		(*Homo sapiens*)	? Buried Channel of Thames Flood Plain Terrace of Thames
Third Glacial	Little Eastern Glacier: Irish Sea Ice	" Cave " Man (*Homo neanderthalensis*) Mousterian	Newer Drift. Cromer Moraine Coombe rock (glacial sludge) formed in south
Second Inter-Glacial	Long Warm Period	implements Acheulian implements	Taplow (50-foot) Thames Terrace, covered with loess or brickearth. Mammoth Swanscombe (100-foot) or Iver-Boyn Thames Terrace with warm water molluscs and southern mammals
Second Glacial	Maximum Glaciation	" River Drift " Man Chellean imple-ments	Ice-sheet nearly as far south as London Deposition of Older Drift (including Chalky Boulder Clay of East Anglia)
First Inter-Glacial			
First Glacial	North Sea (Scandinavian) Ice-Sheet	Pre-Chellean	Norwich Brickearth of north-east Norfolk
Pre-Glacial			Pre-Glacial Raised Beach of the south-west

THE NEEDLES, ISLE OF WIGHT, LOOKING EAST AERO-PICTORIAL

These spectacular rocks consist of chalk, folded by Alpine earth movements so as to be nearly vertical. The dip of the rocks decreases to the south (on the right); to the north in Alum Bay the beds which succeed the chalk have been folded so that they are truly vertical

PORTLAND BILL, DORSET (PORTLAND LIMESTONE) L. D. STAMP
Illustrating the undercutting action of the sea, by which an almost flat peneplane of marine erosion is produced

H.M. GEOLOGICAL SURVEY
SOUTH-WEST OF ST. DAVID'S, PEMBROKESHIRE
A plane of marine erosion of Pliocene age, with hills which were once islands

when the land was *above* its present level and deep channels were excavated there was a renewed lowering of the land surface and a final glaciation. Man retreated before its advance but continued to live in caves such as Creswell outside the York moraine which marks the limit there of the new ice advance. Man survived in East Anglia beyond the ice-limit, which was about Hunstanton. This was the period of Magdalenian man or the Creswellian culture. To the west of the Pennines the Irish Sea ice again invaded parts of the Cheshire Plain but this time did not penetrate as far south as the South Shropshire Hills. Man, however, was driven from the Welsh caves where the remains of Aurignacian man were sealed up. This last glacial period was one when the last Coombe Rock was formed in the south ; the late glacial deposits with their interesting plant remains were formed in the Lea Valley (Ponders End Stage).[1] In the surface features which it has left in the north of England, it is the most important of the glaciations from the point of view of scenery. The severe last glaciation of the Lake District is attributed to this period.

As recently as 1932 Boswell was content to end the geological story at this stage, remarking, " The passing of the Ice Age was marked by a slow but steady subsidence of the land area of southern and eastern England, which commenced apparently after late Aurignacian times and continued until after the Neolithic period. Thus were produced the submerged forests, the drowned river valleys and the buried valleys which occur just below the present-day flood-plain deposits of the rivers."

It is typical of the fascinating fields of knowledge still to be explored that the new methods of pollen-analysis of successive layers of peat have enabled a whole history of change to be sketched, covering the period from the retreat of the last ice-sheet to the present day, though the interpretation is still full of pitfalls.

The geography of the north and Midlands of England at a late stage in the glaciation of the country is depicted in Fig. 61. This map may be regarded as generalised : thus the conditions on the whole would be very similar at the end of the Third Glaciation (the Little Eastern Glacier, marked on the map, is of the Third Glaciation) and at the end of the Fourth when the final retreat began. The essential

[1] Certain botanists have pointed out that the designation " arctic " is something of an exaggeration. The Ponder's End flora is a colder one than flourishes in the neighbourhood to-day but would not be out of place in the north of Scotland.

point is that the ice-sheets blocked the northward flow of rivers : their waters were ponded back and formed extensive lakes to which the melting ice-sheets added a large volume of water.

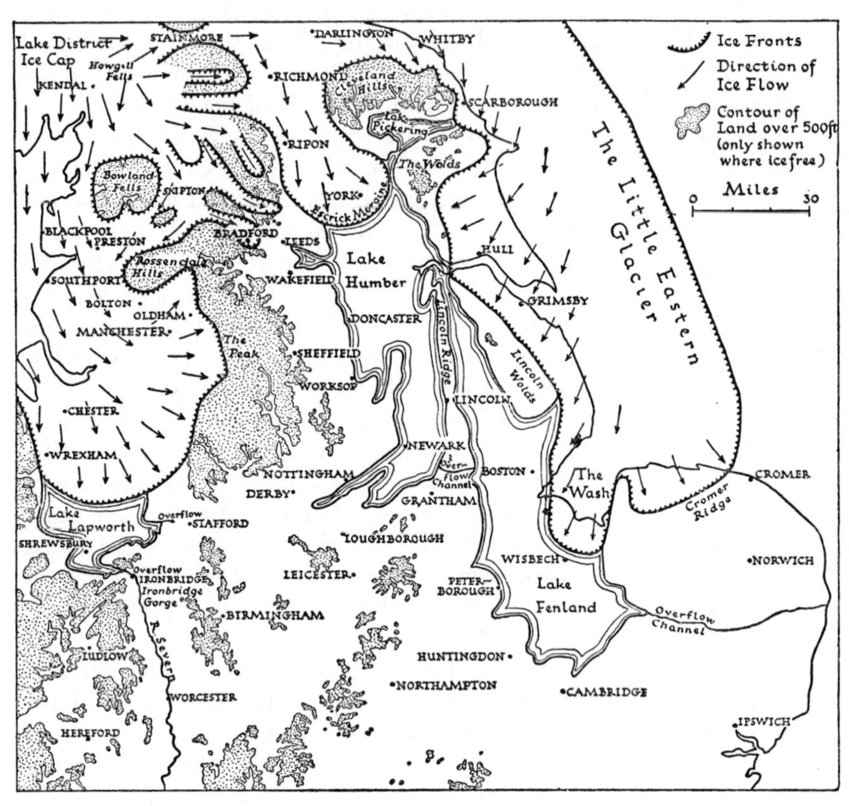

Fig. 61.—Glacial Lakes in the Midlands and North of England

The map shows how the waters of the Yorkshire Ouse and the Trent were prevented by the ice barrier from escaping via the Humber to the North Sea and how their imprisoned waters formed a great lake which has been called Lake Humber. At the same time the rivers normally flowing to the Wash were also held up and another large lake, coinciding roughly in extent with Fenland, resulted. At least for a time these two lakes were connected by an overflow channel, now marked by the dry Ancaster gap west of Sleaford, and it is probable that their combined waters escaped to the southern North Sea by the Waveney valley. On the other side of the English Midlands the waters of the Dee were prevented from escaping to the Irish Sea and were ponded back to form a great lake named Lake Lapworth, in honour of the late Professor Charles Lapworth of the University of Birmingham who was for long a leading figure in geological research. As the waters in this lake reached their highest levels they sought a line of escape to the south. The great volume of pent-up waters was doubtless responsible for cutting the famous gorge of the Severn through Ironbridge and there was thus instituted the river's curious semi-circular course. In these glacial lakes a variety of material was water-sorted and deposited —there are sometimes shore-gravels, but sands (frequently giving very light soils) and well-bedded silts (affording some high quality soils) predominate. As the ice retreated the pre-existing drainage in some cases was restored (as with the Humber) but in other cases, as with the Severn, the change resulting from glacial interference remained permanent. The fascinating case of Lake Pickering is discussed later. As the ice disappeared the lakes dwindled and eventually nothing remained but some extensive swamps. We may instance the Fens themselves, whilst the last remnants of Lake Humber may be seen in the swamps of Hatfield Moors and Thorne Moors.

In these glacial lakes the silts or clays are often minutely banded since the more rapid melting of the ice in summer caused slightly coarser material to be deposited than in the winter. The varves or double bands—the coarse representing the summer and the fine the winter—thus represent each a year, and by counting the bands in these varved sediments it is possible to determine exactly in years the duration of the period since the end of the Ice Age. A post-Glacial chronology on this basis was worked out in great detail in Sweden by the geologist the late Baron de Geer. He was also able to date exactly some of the terminal moraines which mark stages in the gradual retreat of the ice, because they separate the basins in which the varved sediments occur.

This Swedish work has been at one and the same time a source of inspiration and a serious pitfall into which many English workers have fallen headlong. At the present day the higher parts of Sweden, though not ice-capped, are in the Arctic-Alpine region and the bulk of the country is in the Coniferous Forest Belt, whereas most of Britain is in the Deciduous Forest Region. Thus when Sweden was still ice-covered the whole of Britain was probably already free of ice and covered with a tundra vegetation, in due course to be succeeded by coniferous forest and then by deciduous forest. Thus whilst the *sequence* of post-Glacial vegetational stages may be the same in England as in Sweden it is wrong to assume contemporaneity of any given stage and the chronological scale in years established in Sweden is not directly applicable in this country.

With the end of the glacial period proper in Britain and the disappearance of certain arctic mammals such as the mammoth, sub-arctic or tundra conditions prevailed and the country was the home of reindeer, Arctic fox, marmot and other tundra animals. A little later some species typical of the steppes were found—the jerboa, steppe marmot and horse—but Britain was almost always too moist for typical steppeland conditions. It is following the tundra-steppe period that the work of Godwin on pollen becomes important. Worked out for the East Anglian Fenland his post-glacial period starts with a *Birch-Pine zone* in which the dominant tree-pollen is birch, with pine and a little willow, but in which the proportion of non-tree pollen is so high as to suggest an open landscape. This is succeeded by a *Pine zone* in which pine definitely replaces birch as the dominant pollen and in which the pollen of *Corylus* (hazel) and various warmth-loving trees begins to become frequent. In the succeeding *Pine-Hazel zone* there are three horizons. In the lowest elm extends swiftly, oak though present is less important ; pine is still dominant and hazel is abundant. In the middle horizon oak becomes equal in abundance or even surpasses elm though hazel remains abundant. In the upper horizon pine becomes less important whereas oak was increasing, but the climate was clearly a dry one and enabled pine to be locally dominant. This pine-hazel zone corresponds with what is called the Boreal Climatic period (of Blyatt and Sernander and other writers) and contains late Tardenoisian implements in Fenland and early Mesolithic implements elsewhere.

Godwin has shown that peat dredged from the floor of the North Sea belongs to several different horizons, including the three just described which he calls IV, V and VI. He has also shown that the well-known submerged forests round the coasts of Britain belong to several horizons and there is not just one Neolithic Submerged Forest. It was formerly assumed that the submerged forests round the shores of Britain, associated with polished stone-axes of Neolithic man, were all of the same age. It is clear, however, that the land had been steadily sinking during Godwin's zones IV, V and VI—the Boreal Period—and as already stated there are reasons for believing that the *sudden* floristic change at the end of the Boreal period was due to the new and final severance of Britain from the Continent, the opening of the Strait of Dover and the establishment of the all-important circum-insular circulation of water with the consequent establishment of an Atlantic type of climate broadly comparable with that of the present day. It is at least quite clear that at the end of Godwin's Zone VI there is a sudden replacement of pine by the moisture-loving alder as the dominant tree pollen. Throughout the south and east of Britain alder and the trees of the mixed oak forest then gradually become dominant. Godwin has tentatively distinguished an *Alder-Oak-Elm-Lime zone* (VII) associated with Neolithic implements, succeeded by a transitional level corresponding with the Bronze and Iron Ages and then by an *Alder-Oak-Elm-Birch* (*Beech*) *zone* (VIII) which is associated with Romano-British remains. Through the long period of the Zones VII and VIII, during which the dominant vegetation of the British Isles was deciduous forest as it is at the present day, there were still certain climatic fluctuations. Widespread remains of forest with Bronze Age relics in the midst of bogs suggest a drier and perhaps colder spell between two wetter. Reference is often made to the climatic "deterioration" which occurred about the time of transition from the Bronze to the Iron Ages (*circa* 500 B.C.) when the prevailingly wet and mild conditions rendered "blanket-bog" the "climatic-climax" vegetation of much of western Scotland and Ireland. The "blanket-bog" is so-called because it covers the whole surface, including low hills, like a blanket and is a "climatic-climax" vegetation because it is the final one, the climax, which develops under the climatic conditions which exist.

With the close of the Neolithic submerged forest period it is probable that the relative level of land and sea did not differ widely from the present position and the outlines of our shores were approximately as at present also. There has since been the silting up of such channels as that between Thanet and the mainland of Kent; the accumulation of alluvium to make up much of the estuarine marshland and the greater part of some large areas such as Romney Marsh and the growth of shingle banks and spits and of areas of sand dunes. Elsewhere there has been loss of land by coastal erosion but with the passing of the Ice Age the study of these matters falls as much into the domain of the pre-historian as into that of the geologist. From the broad point of view of its geological structure, the Ice Age put the finishing touches, very important it is true, to the building of the British Isles.

Since reference has been made to the chronological scale worked out for Sweden, the following table is given for reference. It is based on the work of O. Pratje, incorporating the researches of De Geer, A. Blyatt and R. Sernander :

	De Geer	*Climatic Periods*	*Baltic Sea Stages*	*Cultural Periods*
		Sub-Atlantic	*Mya* Sea	Historic Times Iron Age
B.C. 1000		Sub-Boreal	*Limnaea* Lake	Bronze Age
2000				
3000	Post-Glacial	Atlantic	*Littorina* Sea	Younger Neolithic
4000				
5000		Boreal		Older Neolithic
6000				
7000 8000	Fini-Glacial	Sub-Arctic	*Ancylus* Lake	Oldest Neolithic
9000 10000 15000	Gothi-Glacial	Arctic	*Yoldia* Sea Ice-covered	
18000	Dani-Glacial			

The *Dani-Glacial* was the period of the ice retreat from Denmark—when ice still covered the whole of Norway and Sweden.

The *Gothi-Glacial* was the period of the retreat of the ice across southern Sweden and lasted from about 15,000 B.C. to 9,000 B.C. The place of the ice in southern Sweden and the Baltic Sea was taken by an arm of the North Sea—the *Yoldia* Sea. The seas are called after their characteristic mollusc.

The *Fini-Glacial* lasted from about 8500 B.C. to 6500 B.C. and was the period of the retreat of the ice from Finland, with a long pause which resulted in the formation of the famous end-moraines (Salpausselka) in Finland. With the removal of the great weight of ice the land surface rose relative to the sea level and the Baltic was converted into a fresh-water lake (*Ancylus* Lake). By inference, this must have been the period when the southern half of the North Sea was land. Around the remaining ice on the high land of Norway and Sweden must have been a belt of tundra ; outwards this gave place to a belt of coniferous forest and it is the remains of peat from this forest which are dredged up from the present floor of the North Sea as moorlog. This, as noted above, Godwin regards as belonging to his pollen zones IV, V and VI, the age of which would thus be 8500 B.C. to 6500 B.C.

The *Post-Glacial* of De Geer thus began about B.C. 6500 and the final retreat of the Scandinavian ice, about 6000 B.C., was accompanied by a very marked and widespread lowering of the land surface and consequent transgression of the waters of the North Sea. They spread into the Baltic as the *Littorina* Sea, over the low ground of Flanders as the Middle Flandrian Transgression of French geologists, and finally severed Britain from the Continent by linking the North Sea and the English Channel. This gives us a date of about 5000 B.C. for the final severance and for the sudden establishment of the Atlantic type of climate, and about 4000 B.C. for the greatest extension of the sea and the formation of the 25-foot Neolithic Raised Beach.

It may perhaps be remarked that one's general reaction to this chronology is that the time must be regarded as a minimum and seems to be somewhat inadequate for the immense amount of denudation which has taken place since, especially in some coastal areas. Whilst it gives us a date for the final severance of Britain from the Continent, it does not give us a date for the final retreat of the ice from the different parts of the country.

In the twenty years since this chapter was first drafted there has been much research on Pleistocene deposits and the history of the great Ice Age. In particular methods have been evolved of dating with considerable precision various of the later deposits. The method of radio-carbon dating is based on the fact that carbon ^{14}, a radioactive isotope of carbon with an atomic weight of 14 created in the upper atmosphere, is absorbed by all living matter—notably in the case of plants by photo-synthesis. After death absorption ceases and content diminishes at a known rate so that age of fossil wood, bones, and other organic remains can be determined by measuring proportion of radio-carbon to total carbon. It is an accurate measure up to 20,000 or even 30,000 years. See also p.153

THE REGIONS OF BRITAIN

The London and Hampshire Basins

THE DISTINCTION between Lowland Britain and Highland Britain was stressed in an early chapter of this book and it has now become abundantly evident that neither of these two fundamental divisions forms a homogeneous whole. That each is made up of distinctive parts is clear from the number of regional names which are in everyday use—we talk naturally, for, example, of the Scottish Highlands, the Lake District, the Yorkshire and Lincolnshire Wolds, the Downlands, Fenland, and East Anglia. There are other parts of the country to which one would refer more naturally by mentioning the counties concerned. It is more probable that one would say Devon and Cornwall rather than the South-Western Peninsula or Cornubia.

The geographer, with the aid of the geologist and geomorphologist, has endeavoured to introduce some degree of consistency into this idea of regions. Thus it is now possible to divide the two halves of Britain each into a number of regions which, because they are self-defined by features of land relief, geology or structure, and climate, and have a resulting homogeneity of vegetational, animal and human response to the environment, are usually referred to as " natural regions." We will not enter into the favourite discussion as to what constitutes a natural region ; we will content ourselves by saying that they are convenient divisions of a country for descriptive purposes and are particularly appropriately used in a work such as the present which seeks to explain the origin of the varied habitats of the native animals and plants, since each natural region is, in effect, a single type of habitat or a group of related habitats.

In the chapters which follow an attempt has accordingly been made, in what is admittedly an impossibly small space, to call attention to a few of the most outstanding physical features of the many contrasted regions of the British Isles and at the same time to arouse interest in the fascinating story that every hill and valley, every river and lake, indeed every fragment of rock, may reveal to the careful

NEAR CLIFDEN, CONNEMARA, IRELAND L. D. STAMP

A glaciated landscape on ancient rocks which outcrop at the surface make the cultivation of all but tiny patches of land impossible

H.M. GEOLOGICAL SURVEY

REMAINS OF A PINE FOREST IN PEAT, NEAR DALESS, NAIRN

Illustrating former climatic conditions which favoured the growth of pines where none now grow

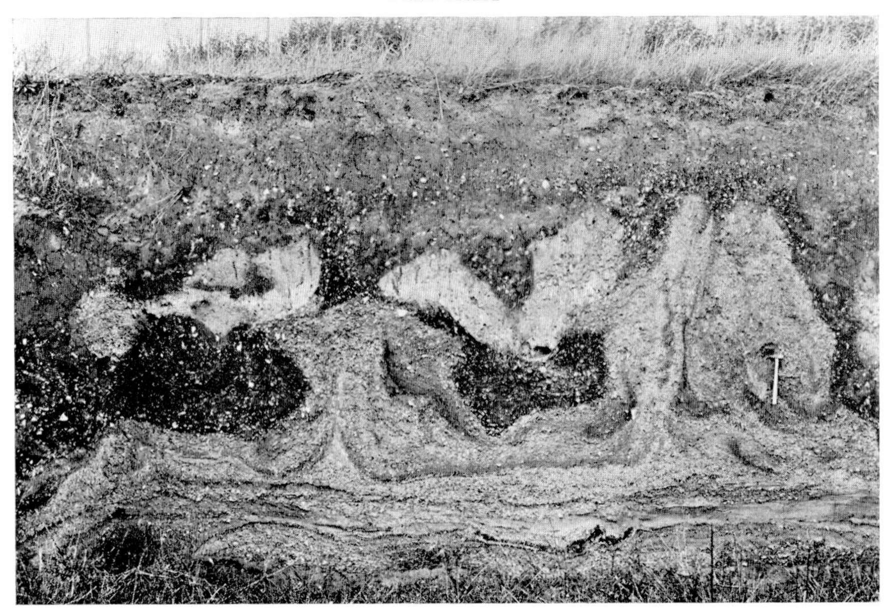

H.M. GEOLOGICAL SURVEY

CONTORTED GRAVELS OF THE RIVER CAM, GREAT SHELFORD, CAMBS.
When great ice-sheets overrode East Anglia, they "rucked" up and so contorted pre-existing sheets of gravel

THE SOUTHERN UPLANDS AEROFILMS
This spectacular air view, taken in the early morning, shows the gently rounded forms of a mature landscape but with deep gullies illustrating rejuvenation by uplift

observer. It would be logical, perhaps, to start with the oldest parts of the country, with the Highlands of Scotland, and to work gradually southwards to territory where the surface features are all young in terms of geological time. But the reverse order has been followed, on the grounds that what is familiar in every day to the majority of readers should be made first to tell its story. The very brief treatment which has had to be accorded to the complex regions of Scotland and Ireland may be compensated for in part by the issue of special volumes in this series dealing with the natural history of the Highlands and of Ireland, where the physical environment will naturally be considered.

The London Basin.

The site of London is one of the supremely satisfying examples of the continued but changing influence—often so subtle as to be completely unrealised—of the factors of the environment on the life of man. Originally twin hills, gravel-capped remnants of the 50-foot terrace, on the north bank of the Thames, afforded dry sites above the reach of floods just where the northward swing of the river brought the main channel along that bank and allowed direct access by boat. The river may have been fordable at low tide but the marshy character of the south bank made a difficult approach from that side and added to the importance of a natural defensive site. So the twin hills were enclosed by a wall and became the Roman Londinium with bounds which are but little different from those of the present City. On the one hill the Royal Exchange and the Bank of England have replaced the Roman Forum, on the other the Cathedral of St. Paul replaces the Temple of Venus, but the little valley of the Wall Brook between the two is now only remembered in the thoroughfare known as Walbrook.

Rising above the floodable marshlands other " islands " of terrace gravels afforded settlement sites and the Saxon suffix " ey " or " ea " (island) is perpetuated in Chelsea, Battersea, Putney and others, though Thorney on which the monks built their Abbey to the west of London (Westminster) is now an almost forgotten name. For a thousand years these scattered " dry-point " settlements remained typical of the heart of what is now Greater London ; the lowest fording point of the Thames became (and still remains) the lowest bridging point of the river and the construction of London Bridge with its narrow arches restricted the port to its present site below the bridge. Naturally a settlement grew up at the southern end of the bridge. The

FIG. 62.—The Regions of Britain

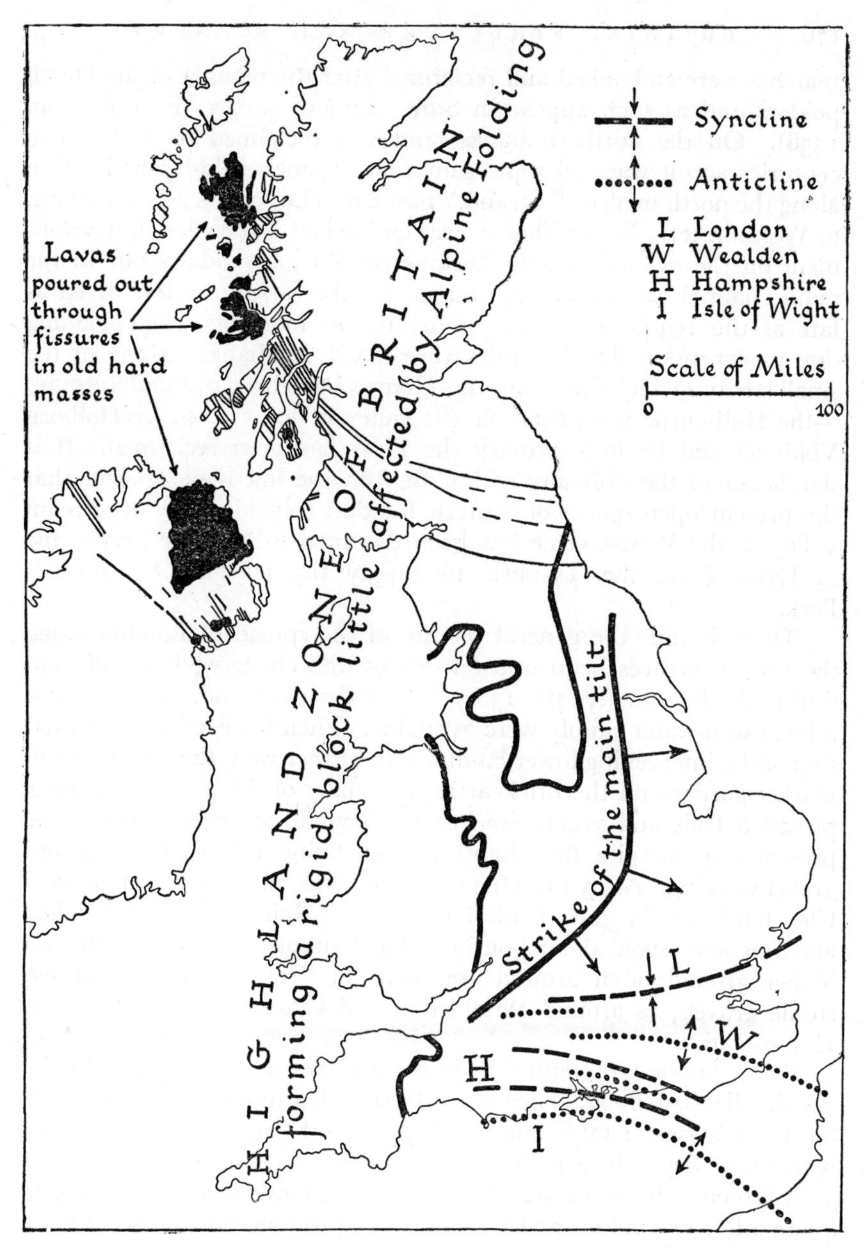

FIG. 63.—The Structural Elements in the geography of Britain
Western Scotland shows the " dyke-swarm " mentioned on p. 235.

marshes were embanked and reclaimed after the manner of the Dutch polders and as such appear in Stow's famous survey (First Edition, 1558). On the northern banks land was reclaimed too : but two centuries ago it was still a pleasant walk through fields from London along the north bank or " Strand," past a number of fine new mansions, to Westminster. As the Thames was embanked and enclosed, development on the old marshlands became possible, but tidal scour in the main channel was increased, and so was the range of tides. Even as late as the 1930's there have been anxious moments when flooding due to exceptionally high tides threatened Millbank. Many of the small streams which drain into the Thames have been put underground —the Holbourne is an example (its valley is that crossed by Holborn Viaduct) and its little estuary the Fleet has been reclaimed. It is significant of the difficulty of building on the low-lying ground that the present open spaces of western London coincide with the stream valleys : the West-bourne has been dammed to feed the Serpentine in Hyde Park, the Ty-burn to supply the lakes in St. James's Park.

There is thus the general picture of the spread of London along the gravel terraces bequeathed to us by the changing levels of land during the Ice Age (see pp. 156-7) where dry sites, and, in early days, a local well-water supply were available. Then followed the reclamation of the intervening lower land. From further west, the orchards and market gardens on the brickearths, especially of the Taplow Terrace, provided fruit and vegetables just as they still struggle to do at the present day, despite the almost overwhelming tide of housing and gravel working. Away from the gravel terraces, the heavy, wet London Clay land was in grass fields for London's dairy cows and for hay and was long avoided for housing. The " suburbs " developed first on higher ground often around commons or " waste " on high-level, sterile gravels, as around Wimbledon and Clapham Commons or at Hampstead.

Below London, on both sides of the Thames, were extensive alluvial marshes liable to flood. They formed tolerable cattle and sheep pastures but the high-water table and liability to flood limited their usefulness. By contrast dry hills near the river afforded valuable settlement sites—as at Greenwich, Woolwich, Erith, Purfleet, Grays and Gravesend, to quote a few examples—and it is interesting to note how many of these owe their good sites to minor folds in the midst of the London Basin

Plate XXIII

THE NORFOLK BROADS NORFOLK AND NORWICH AERO CLUB

These remarkable shallow stretches of water occupy deserted mediaeval peat-workings. Wroxham Broad is in the foreground; Hoveton Broad on the left, Salhouse Great Broad on the right

Plate XXIV

SALT MARSHES ON THE ESSEX COAST

AEROFILMS

Illustrating the development of salt marshes and their gradual reclamation (shown in the foreground) by a retaining "seawall"

syncline which brings the chalk to the surface. In due course when the right time came—as it did during last century—the alluvial lands below London proved the ideal sites for the easily excavated docks now making up the largest dock area of any port in the world.

It has already been made clear that the lower Thames flows roughly along the central trough of the London Basin syncline. Though the basin was initiated in earliest Tertiary times, if not before, and was an original basin of deposition, its present form is mainly the result of Alpine folding and, true to the general form of Alpine folds, its southern limb, represented by the Chalk of the North Downs, is more steeply inclined than the northern limb represented by the Chalk of the Chilterns. The fold is thus slightly asymmetric and so is the valley. The river, like the hypothetical river in Fig. 30, flows from west to east near the centre of the syncline. Apart from minor folds parallel to the main one, there are others with axes from north to south or from north-west to south-east or south-west to north-east. These minor folds play a part in determining the remarkable variety within the London basin : it has been claimed for example that the course of the north-south Lea Valley and the trend of the Epping Forest ridge are directly due to north-south folding.

The general picture of the London Basin is thus first of diverging areas of chalk uplands—the Hog's Back and North Downs from west to east on the southern side and the Chilterns from south-west to north-east on the northern side. These two arms come together in the west of Berkshire and the Thames breaks into the basin through the northern arm by the beautiful Goring Gap. This gap runs roughly from north-west to south-east and probably originated as described on p. 201. The scenery of the chalklands is diversified by the many superficial deposits which mask the chalk itself—by the residual Clay-with-Flints (see p. 76) with its heavy soils, lush meadows and damp oak woodland, by the high-level gravels (see p. 150) of Pliocene and later ages with their hungry sterile soils and oak-birch heaths, by the boulder clays of the northern side or Coombe Rock of the southern with varied agricultural soils. The chalk itself may afford undulating stretches of arable in large fields, or wide areas of short grass (fescue downland) where the soils are too thin to plough or the land too steep, merging in the steepest areas into juniper scrub or replaced by beech " hangers." Woodland of beech, pine or larch, locally with yew, occur elsewhere on the chalk itself. Some of the different types of chalk scenery are illustrated in

Plates 21 and 23. Plate 21 shows a typical chalk scarp with fescue grasses and yew, Plate 23 is a cultivated dry chalk valley.

The rocks which normally succeed the Chalk are the lower beds of the Eocene—first the varied group known locally as the Lower London Tertiaries and then the main mass of the London Clay which is succeeded by the sandy Bagshot Beds. The Lower London Tertiaries are predominantly "continental"—consisting of fluviatile sands, mottled clays and gravels—in the west of the basin and predominantly marine or estuarine sands, clays and pebble beds in the east.

The detailed sequence is best expressed thus :

	West and North	*North-east Kent*	*East Kent*
Ypresian	Claygate Beds		
	London Clay	London Clay	London Clay
	London Clay Basement Beds		
	—	Blackheath Pebble Beds	Oldhaven Sands
Landenian	Reading Beds (mottled clays and sands)	Woolwich Beds (estuarine clays and sands)	Woolwich Beds (marine sands)
	absent	Thanet Sands (marine)	Thanet Sands (marine)

Reference has already been made to the interesting Bullhead Bed at the base of the Thanet Sands (p. 140). The fine-grained or loamy Thanet Sands give rise to some excellent, loamy, well-drained soils and form a large part of the Fruit Belt of north-west Kent, to which the shelly clays of the Woolwich beds also contribute.

The Blackheath Pebble Beds with their remarkably well-rounded jet-black pebbles, mainly of flint, give rise, on the other hand, to sterile hungry soils and despite their unconsolidated nature seem in some way to be resistant to weathering. Thus the overlying London Clay has been swept away by denudation, leaving the Blackheath Pebble Beds to cap such considerable plateau surfaces as Chislehurst and St. Paul's Cray Commons and the Addington Hills as well as Blackheath itself.

The London Clay is the dominant Eocene rock of the London Basin and is associated with rather featureless topography such as that of

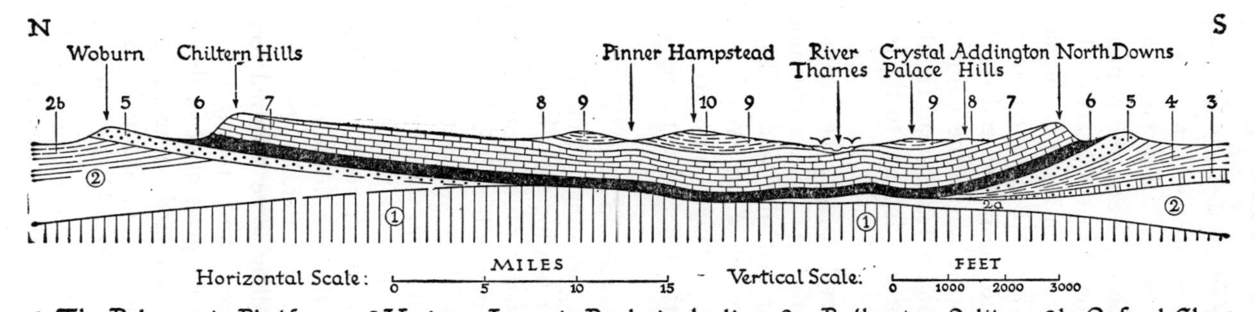

FIG. 64.—Section across the London Basin. Reference is made on p. 180 to the artesian basin under London. Rain falling on the chalklands of the Chiltern Hills and North Downs sinks into the ground and flows underground towards the centre of the basin. The impervious Gault (shown in black) acts as a floor to the basin which is thus like a gigantic pie-dish. When the "crust," formed by the London Clay is pierced by a well the water, if under sufficient pressure, will gush out of the surface because the level of the water table in the Chilterns and North Downs is higher than the surface level at London

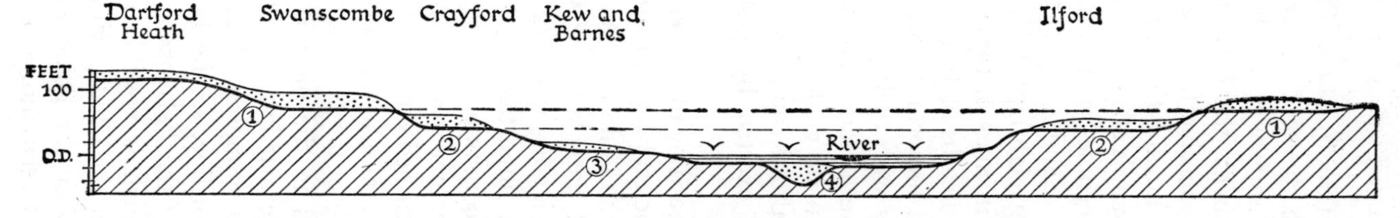

FIG. 65.—Section across the Thames Valley showing River Terraces

much of South Essex. The stiffest clays have in places defied cultivation through the ages and remain in woodland—such as the Blean Woods near Canterbury or the Ashtead Woods of Surrey. The latter are interesting ; the Romans tried cultivating the clays and the remains of an extensive farm or " villa " are to be found in the heart of the woods, but for the last 1700 or 1800 years the natural soil factor has defied man's efforts. In the upper part (the Claygate Beds) the London Clay becomes more sandy and passes up into the coarse and sterile Bagshot Sands. Apart from the capping of such hills as the Laindon and Rayleigh Hills in Essex and the high ground of Hampstead Heath, the main mass of these sands is in Surrey, there giving rise to the extensive heaths of Bagshot and its neighbours.

The important part played by the high-level gravels, the Thames terraces with their gravels and brickearths, and the Thames alluvium we have already considered. It will be clear that the London Basin rivals in its geological interest the classic ground of the Weald and both afford outstanding examples of the detailed influence or control exerted by geological structure on the life and activities of man.

Regarding the underground structure of the London Basin (Fig. 64), the synclinal form of the rocks down to and including the chalk has given rise to an artesian basin, the chief water-bearing rocks being the Thanet Sand and the Chalk. Though the latter varies greatly in its water-yielding qualities according to the extent to which it is fissured, it is an important source of supply to private deep wells rather than to the communal supply which is drawn chiefly from the Thames—to a less extent from the Lea. When artesian wells were first bored the water gushed out with considerable force at the surface and at one time the fountains in Trafalgar Square " played " through this force alone. The use of large quantities of this stored artesian water, however, has resulted in the natural level of the water falling some hundred feet and the artesian wells have become sub-artesian and no longer flow without pumping. Just as the Wealden anticline gives place in depth to a syncline, so the London syncline is superimposed on an ancient ridge of the Palaeozoic platform which is not only an old uplift axis but which is possible itself anticlinal in structure. Under parts of London the Chalk, with only a thin representative of the Gault, rests directly on the Palaeozoics : elsewhere are thin representatives of the Lower Greensand.

L. D. STAMP

The chalk scarp of the North Downs, looking eastwards from near Box Hill, Surrey (See page 178 and page 187 and Fig. 27)

Iping Common, near Midhurst, Sussex T. F. MACKAY

... developed on Lower Greensand. In the distance, southwards, can be

The Hampshire Basin.

The once continuous Tertiary basin of deposition is now separated into the two main basins of London and Hampshire by the western extension of the Wealden uplift. Over this area—northern Hampshire—Chalk outcrops over a wide area and all trace of Tertiary beds has disappeared. There are minor " puckers " of Alpine age and form—such as the Pewsey anticline, a miniature Weald, from the centre of which denudation has removed all beds down to the Upper Greensand—but the main fold is the extensive syncline which forms the Hampshire basin proper. Like the London basin, it is asymmetrical with a southern limb steeper than the northern but the fold was nearer the heart of the Alpine storm and the asymmetry is accentuated, the southern limb being vertical along most of its length (see p. 68 and Plate XIX).

As in the western part of the London basin the Thanet Sands are absent and the Reading beds unimportant. The London Clay occupies a considerable area and gives rise to wet land with heavy soils, exemplified by the Forest of Bere. The later Tertiary deposits of the Hampshire Basin differ from those of London in the thickness and variety of the beds which succeed the London Clay, with a full sequence of Oligocene. Though there are clayey horizons, sandy beds predominate and the extensive area of sterile soils which results is responsible for the New Forest. In another obvious way the Hampshire basin differs from the London : its heart has been invaded by the sea. Not only does the sea penetrate from the eastern end but it has breached the southern chalk wall between the Needles in the Isle of Wight and Ballard Point in Dorset and the two arms of the Solent and Spithead join and unite with the drowned valley of Southampton Water. There are gravels at differing levels as there are in the London Basin but their effect is less marked because of the already wide extent of the sands and gravels of Eocene age.

The Isle of Wight may be described as a geological museum. Its structure is well displayed in the numerous cliff sections and some of its features have become world famous. There is a central ridge of vertical chalk ; to the north is part of the Eocene-Oligocene basin ; to the south an asymmetric anticline exposing rocks down to the Atherfield Clay. The lower and middle Eocene beds, folded with the Chalk, have given rise to the famous section of Alum Bay where coloured sands and clays stand in vertical bands (see Plate XIX). The

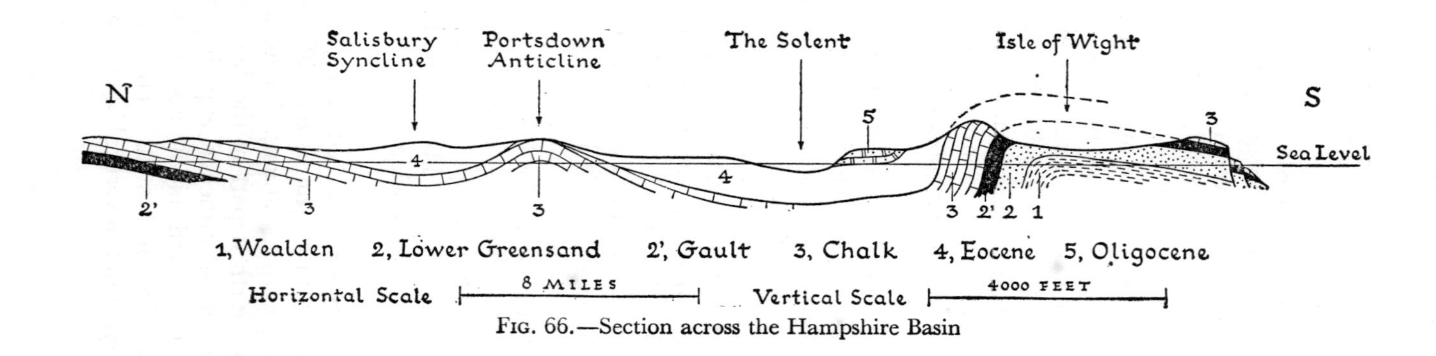

FIG. 66.—Section across the Hampshire Basin

higher Eocene and Oligocene beds include sands and clays with some thin limestone bands—some of them cream-coloured limestones with freshwater shells inviting comparison with the similar limestones much better developed in the Paris Basin.

The structure of the Isle of Purbeck is broadly similar, except that older rocks are exposed to the south. There the central chalk ridge has a double gap near the centre with the ruins of Corfe Castle on the central knoll.

CHAPTER 16

THE WEALD

LONDON is extraordinarily fortunate in having, virtually on its southern doorstep, one of the most fascinating geological regions of all Europe. The area long known as the Weald is properly the oval-shaped tract bounded by the inward facing scarps of the chalk rim and truncated at its eastern end by the Strait of Dover. This is the area originally covered by the primeval forest of Andreas Wald (cf. A. S. *wald*, mod. Germ. *wald*) from which the Weald derives its name. The central mass of sandstone hills, still largely wooded and culminating in Forest Ridges, is often distinguished as the High Weald, the surrounding ring of lowlands and plains as the Low Weald, though local inhabitants tend to think only of these low and still partially wooded claylands when they refer to the Weald of Kent or of Surrey.

Thus defined, the Weald occupies the south-western part of Kent, the southern half of Surrey and the northern half of Sussex and extends just across the county border into Hampshire. Geologically it is difficult to consider the Weald proper apart from its chalk rim and if the North and South Downs are included the unit becomes broadly the three counties of Kent, Surrey and Sussex—the south-eastern peninsula of England lying south of the Thames.

The Weald well illustrates one of the fascinating features of British geology. Despite its proximity to the capital, despite its familiarity to successive generations of geologists, its use as a training ground for students, and its provision of the text book examples of geological phenomena, it is slow to yield up all its secrets. There exists a standing committee, the Weald Research Committee, of that strong society of geologists, largely amateur, known as the Geologists' Association, and this in itself is the recognition of problems still to be solved. It was for some years regarded as the Weald which yielded the much-discussed remains of the near-human *Eoanthropus dawsoni*—a triumphal rescue from oblivion by the watchful amateur in the person of a local doctor only to be revealed later as an elaborate hoax, the perpetrator of which will never be known. The Wealden rivers, so exhaustively studied by W. M. Davis and the prototypes of the now universally accepted classification of rivers, have been examined afresh

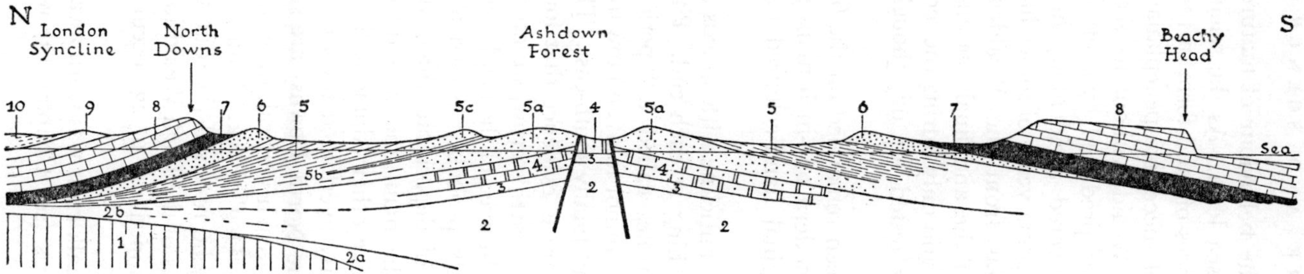

1, Palæozoic Platform. 2, Various Jurassic Rocks, including 2a, Lias and 2b, Bathonian, also 3, Portlandian
4, Purbeckian 5, Wealden, including 5a, Ashdown Sands, 5b, Wadhurst Clay, 5c, Tunbridge Wells Sand
6, Lower Greensand 7, Gault 8, Chalk 9, Lower London Tertiaries (Landenian) 10, London Clay

Horizontal Scale 10 MILES Vertical Scale 4000 FEET

FIG. 67.—Section across the Weald of Kent and Sussex

The scenery of the area is illustrated in Plates XXI—the scarp of the North Downs—and XXII—a typical common on the Lower Greensand.

and in detail to yield evidence of the behaviour of mature rivers to small changes in relative land and sea level. As the science of geomorphology advances, subtle differences of land form in the Wealden hills are made to unfold the story of successive peneplanations. Nor is the underground geology of the Weald without its secrets. The completely hidden Kent coalfield was predicted by geologists, scornfully rejected by experts, then discovered accidentally in 1890 and finally exploited systematically after some years of somewhat doubtful financial manipulation. Natural gas from the Wealden dome, a prototype structurally of so many of the anticlinal gas and oil-fields of the world, was used a century ago and only during the second world war was the structure exhaustively tested for oil, though with no success.

Only parts of the Weald have been remapped on the 6-inch scale by the Geological Survey and the modern one-inch maps published : for the rest there are only the original hand-coloured maps of the survey, made a hundred years ago.

Structurally the Weald is a great anticline with an axis trending in the main from west to east and pitching at each end. Pitching thus at each end the structure is sometimes called a "pericline" (see Plate IIIA). In more detail, there are minor folds and towards the easts the axis trends more south of east and finally south-east. The fold was initiated as an anticline (though there is evidence that for very long it had been an unstable region of the earth's crust) in late Cretaceous times—after the deposition of the main mass of the chalk—and in early Tertiary times the crest of the chalk probably formed a low island incompletely separating the basins of deposition which are now the London and Hampshire basins. The main and subsequent uplift of the fold is doubtless Miocene and like other Alpine folds in southern England the wealden uplift is slightly overfolded towards the north : the dips along the northern rim (the North Downs) are steeper than along the southern and may locally reach vertical, even accompanied by reverse or thrust faulting, in the Hog's Back. A number of strike faults (see p. 23), some of which may be reversed (i.e. due to pressure) are found among the sandstone rocks of the High Weald in contrast to the normal faults which, as explained on p. 23, occur in regions of tension.

At least by Miocene times the chalk crest was worn away, exposing the older rocks ; early Pliocene seas (see Chapter 13) were cutting

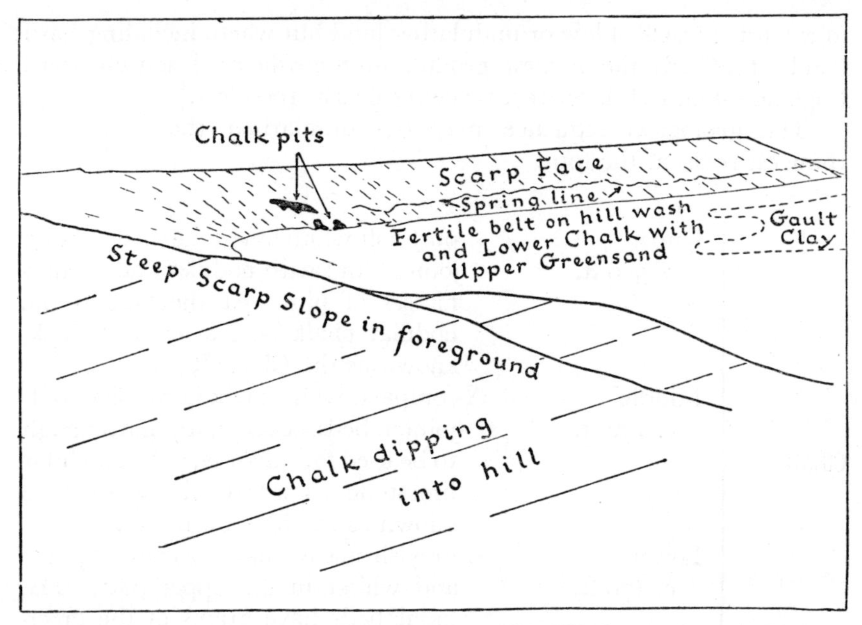

FIG. 68.—Diagram of the Scarp of the Downs near Box Hill, Surrey, shown in Plate 21

platforms or ledges on the chalk slopes and later Pliocene seas probably occupied part of the heart of the Weald, though the main features of its relief are the result of sub-aerial denudation. The pattern of rivers which had been established as a consequence of uplift has been much modified by river-capture, and also by the cutting of the Strait of Dover, but the region escaped direct glaciation.

There is, as a result, a remarkably close correlation between physical features and soils on the one hand and the structure and lithology of the solid rocks on the other. Since the beds below the chalk are chiefly sands or sandstones and clays in alternating sequence they give rise to a succession of ridges or cuestas (see p. 27) and valleys, all the cuestas having scarps facing inwards. Not only, however, are there local variations in the rock sequence within the Weald itself, but local differences in resistance of a given horizon to atmospheric weathering, so that it is both difficult and dangerous to generalise. Thus the Lower Greensand affords some of the most fertile as well as some of the most sterile soils in England : its soft sands may give rise to extensive tracts of low or undulating land but where including hard bands it affords the highest ground in the whole of south-eastern England—Leith Hill, Surrey, reaching nearly 1000 feet.

The geological sequence may be summarised thus—with the youngest rocks at the top :

Chalk	Upper c. 550 ft.	A pure soft white limestone, bedded and jointed and with nodules and tabular masses of flint. At the base is the nodular chalk bed, a few feet thick, known as the Chalk Rock.
	Middle c. 170 ft.	A compact, white chalk in massive well-jointed beds, occasionally hard enough to be used for building, with a nodular limestone a few feet thick at the base known as the Melbourn Rock.
	Lower c. 170 ft.	A greyish marly chalk, becoming purer and whiter in the upper part. The basal beds have grains of the green mineral glauconite.
Upper Greensand	from 0 ft. in the east to 80 ft. in the west.	Sand, sandstone, malmstone and clay, frequently with grains of glauconite. Malmstone is a siliceous rock with a large proportion of sponge spicules : a hard fine-grained variety has been quarried as firestone and used for building : a soft greenish-grey sandstone has been quarried especially near Reigate and Betchworth as hearthstone (used for whitening doorsteps and windowsills).
Gault 100 to 300 ft.	Upper	Marly clay (with a band of phosphatic nodules at the base) passing upwards into marls and siltstones and imperceptibly into the Upper Greensand.
	Lower	Stiff dark clays.

L. D. STAMP

A typical dry chalk valley. Tappington Farm, near Denton, Kent (See page 76 and page 178)

L. D. STAMP

The scarp of the Cotswolds, near Birdlip Hill, Gloucestershire (See page 75 and page 202)

Lower Greensand	Folkestone Beds 60-250 ft.		Sands, generally soft and current-bedded and stained brown or yellow by iron oxides, sometimes with lenticular bands of calcareous sandstone or with irregular seams of carstone (ferruginous sandstone).
	Sandgate Beds 5-120 ft.		Clays, sands and calcareous sandstone with local beds of fuller's earth and, in the west, sandy limestone (Bargate Beds) used for building.
	Hythe Beds 60-350 ft.		Sands (including grey sands or hassock), sandstone, sandy limestone (including Kentish Rag) and cherts.
	Atherfield Clay 15-50 ft.		Blue, brown, reddish and mottled clays, sometimes sandy and with local ironstones.
Wealden Series	Weald Clay		Clays and shales with occasional thin bands of freshwater limestone (Sussex marble)
	Hastings Beds	Tunbridge Wells Sand 130-400 ft.	Coarse sands, locally consolidated into friable sandrock or massive sandstone and with clay bands.
		Wadhurst Clay 100-230 ft.	Alternating clays and shales with bands of sandstone and iron ore and calcareous sandstone (Tilgate Stone).
		Ashdown Beds 160-700 ft.	Sands and sandrock.
		Fairlight Clays	Grey and variegated shales and clays.
Purbeck Beds			Clays, shales and limestones.

The sequence of rocks has been given in some detail because it is important to emphasise the wide differences which are possible from one locality within the Weald to another. Taken as a whole the Hastings Beds are sandstones resistant to weathering and hence their outcrop coincides closely with the High Weald and the sterile soils they yield account for considerable stretches of heathland as well as the woodland, with many Scots pines, of the Ashdown and other forests. Locally, hardening of the sandstones has had some interesting results —the famous High Rocks of Tunbridge Wells form a fault scarp on one side of a valley, whereas the valley of Rocks Wood (near West Hoathly) owes its gorge-like character to erosion along vertical joints in the sandstone. Sometimes water and wind erosion may remove all the softer material and leave isolated blocks (such as the Toad Rock on Rusthall Common) which at first sight recall glacial erratics.

Around the High Weald, it is the outcrop of the Weald Clay and Atherfield Clay which gives rise to the broad plain of the Low Weald or Weald proper. Where the clay is covered by alluvium there are some remarkably fertile stretches and the fruit orchards of mid-Kent have spread from the Greensand soils across this plain (a bad location from the point of view of frost) on to the High Weald.

Where the Lower Greensand sequence has marked hard beds it gives rise to an important scarp or scarps but the features so formed tend to die out. Thus the well-known Leith Hill scarp of Surrey is due to chert[1] beds at the top of the Hythe Beds. The dip slope coincides very closely indeed to the dip of these protective " cap " rocks. In Kent, the calcareous sandstone known as Kentish Rag occurs in beds separated by the soft sands known as hassock and is well developed round Maidstone. Like the Leith Hill cherts it forms a conspicuous scarp, but unlike the cherts the ragstone breaks down into excellent soils. The absence of a marked feature along the Lower Greensand scarp in some areas—for example between Dorking and Reigate—is sometimes due to an absence of hard beds, but sometimes partly to faulting. An unusual feature in the Greensand country of Surrey is the Devil's Punchbowl on Hindhead round the head of which curves the London-Portsmouth Road—it is due to the undermining action of springs at a valley head. Along the southern side of the Weald the scarp due to the Lower Greensand is generally less well marked owing to the absence of hard beds.

Between the Greensand cuesta and the scarp-face of the Chalk of

[1] Chert is a hard flint-like rock, originally a sand but in which the sand grains have been cemented together by the deposition of silica from solution.

the North Downs there is usually a narrow valley known in parts of Kent and Surrey somewhat redundantly as the Vale of Homesdale (or Holmesdale) or, more simply, as Homesdale. It coincides broadly with the outcrop of that dark clay, the Gault, though the southern part may be partly on the soft Folkestone Sands and the northern part is obscured by hill wash from the Chalk. Along the northern face of the South Downs the Lower Chalk tends to form a step or terrace, so that the change from the Downs to the Gault vale is less abrupt and the latter appears to be less marked.

Not only is the Upper Greensand thin ; in the eastern part of the Weald it is rarely seen and makes no separate feature. At the western end of the Weald, the Malmstone gives rise to a cuesta of its own and it was here that Gilbert White noted and described a feature typical also of the Lower Greensand—the sunken roads which are the result of centuries of wheeled traffic powdering the rock and permitting it to be removed by wind and rain. These sunken roads are typical of the loess covered chalk lands across the Channel in northern France : they occur locally and less fully developed in the chalk lands of south-eastern England.

The evolution of the Wealden rivers has been described (page 45). The chalk rim of the Weald is cut at intervals by gaps, but little above sea level, through which the more successful rivers cut. Dry valleys, once occupied by the less successful rivers, are typical of the chalk dip slopes but they were eroded before the scarps had been cut back to their present position probably when the water table was higher. Some, like the Devil's Dyke near Brighton, are so striking that they have come to be included amongst the famous spots of the country. Sometimes dry valleys are now seen to break the line of the scarp and form " wind gaps " whilst it is likely that the valleys or coombes were deepened by the movement of sludge during the Ice Age and it is in them that the Coombe Rock is often found.

Some of the Wealden rivers have features due to recent movements of elevation or depression. They have extensive terraces whilst " knick points " bear witness to recent rejuvenations. The course of the Mole through its gap, between Dorking and Leatherhead, is famous for the numerous swallow holes into which much of the water disappears in dry seasons.

Both strike and dip faults are found in the rocks of the Weald and cause a repetition or a shifting of surface features. In some areas the conditions are such as to cause slipping or slip faulting and this is especially the case where chalk rests on the greasy surface of seaward dipping Gault as in the Warren between Folkestone and Dover. This large unstable area with its odd pools, swamps and piles of debris affords a haven to many a rare plant.

Since this book was first published an exhaustive and authoritative account of the Weald has been written by Professor S. W. Wooldridge (with magnificent illustrations by Frederick Goldring) and published in the *New Naturalist* series.

EAST ANGLIA AND THE FENS

IT IS in East Anglia that one notices the greatest differences between the "solid" and "drift" editions of the geological maps—in other words it is a part of Britain where the surface deposits left by the glaciers of the Ice Age almost completely obscure the solid rocks below.

Unlike some of the names which have been coined to designate regions of the country, East Anglia is a well-known and commonly used name, though not always used with the same connotation. It is applied properly to the old kingdom of the East Angles which, because of its considerable extent, was early separated into the territory of the North Folk and the territory of the South Folk—perpetuated in the names Norfolk and Suffolk respectively. East Anglia is, broadly, Norfolk and Suffolk with the northern part of Essex. Even into mediaeval times it was an isolated area, cut off from the west by the marshy wastes of the undrained fens, bounded on the south by the dense oak forests which for long clothed the clay lowlands of mid-Essex, and girt by the sea on north and east. It was approached on the landward side with relative ease only along the line of the chalk hills and over Newmarket Heath. It is here that a triple line of defence works shows the steps which the East Anglians took to guard themselves from invasion from this direction. The reclamation of the Fens and the disappearance of the Essex forests have broken down something of the isolation but they have not destroyed the essential unity of East Anglia. It forms a distinctive and well-marked region of the country which many find fascinating, though others consider dull.

Regarded as a whole it is essentially a low plateau, nowhere reaching more than 300 or 400 feet above sea level, and in general sloping gently from west to east. Over the western half this low plateau has the white Upper Chalk as its underlying rock. The chalk is dipping eastwards more steeply than the surface slopes so that the zones of the chalk are bevelled off and the highest zones are to be found towards the east where the chalk plunges below a mantle of London Clay (towards the south) or Pliocene "crags." But these "solid" rocks are only to be seen along the sides of the deeper valleys or towards the west in such areas as Newmarket Heath where the drift cover is thin or absent as it often is on higher ground. Although this western margin is a continuation of the chalk scarp of the Chilterns and, with a change of direction, is continued across the Wash as the Lincoln Wolds, there is scarcely anywhere a distinct scarp. If one ever existed, it has been planed down by the passage of ice-sheets. North of Downham Market a bed of Gault or beds of that age appear from beneath the chalk and below that the Lower Greensand which gives rise to a considerable sandy belt well developed round Sandringham.

The crest of the low plateau lies somewhat west of a central line so that the larger rivers of Norfolk—the Bure, Wensum and Yare—flow eastwards to the North Sea ; the Nar and Wissey have shorter courses to the west. The plateau is virtually cut into two halves by the channel of the Little Ouse and Waveney—coinciding with the boundary between Norfolk and Suffolk—and this depression probably marks an overflow channel of glacial times (see p. 167).

If the disposition of the Cretaceous and Tertiary rocks exposed in the valleys is simple, the same certainly cannot be said of the glacial deposits of the surface. The first advance of ice was from the east, from the North Sea, and this North Sea glaciation resulted in the deposition of a loamy deposit over the north-western region between Sheringham, Norwich and Yarmouth. The loams were later overridden by other ice-sheets which contorted or "rucked up" the deposit so that it has been given the name of " Contorted Drift "—but its loamy character results in some magnificent loamy soils. Most of the country here is below the 50-foot contour, flat and featureless except for the shallow valleys in which lie the Broads. An example of contorted gravel is shown in Plate XXII.

The Second or Great Eastern Glacier was very extensive, overrode the country from the north-west, deposited the Chalky Boulder Clay, usually a sticky boulder clay but redeemed in many areas by numerous pebbles and grains of chalk. It is variable in thickness but may reach 100 to 150 feet or more and covers all " High Norfolk " and " High Suffolk." During the retreat it left occasional lakes in which banded clays and loams were deposited, as well as numerous fans of sand and gravel.

The Third or Little Eastern Glacier seems to have advanced from the north but only covered the north of the county and left as a terminal moraine the well-marked Cromer Ridge of coarse sandy and gravelly deposits. The ridge reaches heights of upwards of 300 feet, forms a water parting and gives rise to sandy heathland, recalling other heathlands of similar origin in northern Germany but unlike anything else in East Anglia. South of the Cromer Ridge is a plain of outwash loams and loamy sands. Blakeney " Downs "—a long winding ridge of coarse (cannon shot) gravels—may be a moraine or a kame marking the edge of an ice-dammed lake or an esker (see p. 87). There are also other gravel ridges.

The Fourth or Hessle Glacier seems to have impinged only on the northern coast of Norfolk.

For the most part the boulder clays and drifts of East Anglia afford fertile soils which are the basis of the most extensive cornlands of Britain. Where they become loamy they are especially fertile: only where they become coarse sands and gravels are they too light and hungry to be fertile and give rise to patches of heath.

One of the most remarkable tracts of East Anglia is Breckland: an area of sandy soils so light that they " blow " if vegetation is removed. Breckland covers some 250 square miles partly in Norfolk and partly in Suffolk. The idea that the the sandy soils were of sand blown up from the shores of the Wash is now discarded in favour of the view that they are sandy boulder clays or glacial sands from which the limestone or chalky particles have been leached out (the soils are thus podsols in various stages of development) and the fine clay particles washed out. Breckland is so named from the " brecks " or " brakes "—intakes of land for temporary cultivation which were formerly made. The heaths which formerly covered much of Breckland are of great interest ecologically as the nearest approach in Britain to steppeland conditions. They are in danger of disappearing completely as Breckland is now largely afforested with conifers—the largest continuous area of afforestation (it was planted by the Forestry Commission) in Britain. Interrupting the surface of Breckland are certain shallow lakes or meres which, whilst probably marking the position of solution hollows in the surface of the chalk below, present some problems regarding the source of their water. The water may seep in as the natural result of surface drainage or may result from fluctuations in the water-table so that the meres only fill when the water-table is high.

The Broads, which are the shallow lake-like extensions of the lower courses of the Ant, Bure, Yare and other rivers, long provided a geomorphological puzzle. Investigations by J. N. Jennings were begun in 1939 and published in 1952[1] but further evidence later came to light[2]. It was thought that the subsidence of the Neolithic Submerged Forest Period gave rise to flooding of the lower courses of the rivers but the drifts of beach material prevented these from becoming saline estuaries and caused freshwater " broads " instead. It is now proved that the Broads are essentially flooded mediaeval peat-workings. All the Broads are of course shallow: they pass on the margins into reed swamps. Sometimes the river courses pass through them but at other times the rivers follow a more direct course independent of the Broads.

Reference to the formation of the Broads leads naturally to a brief mention of the extraordinarily fascinating East Anglian coast. There are stretches where normal sea erosion is eating into low cliffs of boulder clay and the rapid regression of the coastline in the neighbourhood of Dunwich is too well known to need description. What might have been the site of the East Anglian cathedral, had not Norwich been chosen, is now part of the North Sea. Elsewhere along the coast there is coastal accretion. This is well seen in the " bars " and "nesses" which obstruct the mouths of the Yare and Alde and cause those rivers to follow a long southward course separated from the sea only by a long narrow spit of sand and gravel. But the classic example is the north coast of Norfolk where a complex series of spits lies on the seaward side of a belt of saltwater marshes seamed by innumerable channels filled at high tide, in turn giving place inland to reclaimed freshwater marshes backed by an old line of cliffs. It must be left to other volumes in this series to describe the varied habitats thus formed but so great is the interest of the area that Scolt Head Island has been accepted by the National Trust for conservation as a nature reserve. The volume on Scolt Head Island, edited by J. A. Steers, shows how intimately geological, geomorphological and ecological studies must be interrelated to permit even a partial solution of the innumerable problems of our coastline. Blakeney Point is shown in Plate X.

Fenland

Except for a few small deliberately or accidentally preserved fragments, the English Fens have ceased to exist, but Fenland remains as

[1] J. M. Jennings: The Origin of the Broads. *Royal Geographical Society Research Series No. 2*, 1952.

[2] E. A. Ellis: *The Broads*, New Naturalist Vol. 46, 1965.

Dartmoor from Hatherleigh Moor, Devon L. D. STAMP

In the foreground are red soils derived from Permian sandstones; in the distance is Dartmoor, consisting of a great mass of granite (See page 78 and page 214)

L. D. STAMP

Carrick Roads, looking south from near Penpoll, Truro
The lower part of the drowned valley shown in the picture below

L. D. STAMP

The Truro River at Malpas, Cornwall
A drowned river valley (See page 60 and page 215)

a unique and fascinating region of the country. No other region has undergone quite such far-reaching changes in historic times—from a waste of swamp and marsh with but occasional inhabited islands to the most continuous fertile stretch of ploughland in Britain. Curiously enough, an intermediate stage is often forgotten—a century ago it was " the vast plain of luscious grass with innumerable fat sheep " which excited the admiration of Cobbett.

A glance at the geological map shows that the western or scarp edge of the Chalk outcrop undergoes several marked changes in strike. The Chiltern edge and its continuation runs north-eastwards : the edge of the chalk outcrop in Norfolk (one can scarcely call it a scarp) is south to north ; the strike of the Lincolnshire Wolds is from south-south-east to north-north-west. These changes in strike may have resulted in regions of tension along which the chalk was shattered or rendered more easily eroded. The first change in strike is marked by the Little Ouse-Waveney depression. The second was, in pre-Glacial times, followed by a main river which drained much of the Midlands of England before escaping to the North Sea, possibly joining first the proto-Rhine. The change in strike has the result of extending the Jurassic clay vale between the main Chalk and Oolite scarps into a broad clay plain, underlain by the great mass of the Kimeridge and Oxford Clays. The Great Eastern Glacier swept across this plain, the ice moving from north-west to south-east, and may have played some part in deepening locally a plain already worn down to base level. In post-Glacial times, with that lowering of the land surface which produced the submerged forests round the coasts of Britain, the sea flowed gently through the erstwhile river gap to cover some 2000 square miles of the Clay Vale. The Wash of the present day is but a small remnant of this British Zuyder Zee. Into the seaward half of the gulf so formed, waves, tides and currents soon began to sweep an accumulation of fine silts ; the rivers, though they brought little sediment, had difficulty in discharging their waters to the North Sea and their waters were ponded back by the accumulation of marine silt. In the stagnant water thus formed fen vegetation became established and the formation of fen peat proceeded apace.

Thus Fenland of to-day consists of four distinct parts. Underlying the whole, though often at great depth, are the Jurassic clays, but only rarely do these " solid " rocks reach the surface. The " islands "—of which the true Isle of Ely is a good example—are more often of boulder clay than of Jurassic clay but in any case with their heavy soils contribute one element in Fenland topography. Around the margins of the former gulf are gravels and sands partaking of the nature of shore deposits and these constitute the second element. That leaves the bulk of Fenland consisting of peats on the landward side and marine silts on the seaward. The silts afford deep, fine-grained, easily worked soils and their value was already appreciated in Romano-British times for they were cultivated in a series of small fields and it is probable that the so-called Roman Bank was actually built to protect these lands from incursion of the sea at abnormally high tides. Thus there was a line of village settlements on the seaward side of the Fens when the landward side was a great waste of waters.

This is not the place to attempt, even in outline, to recapitulate the story of the draining of the Fens. It has been considered in detail by Professor H. C. Darby.[1] To-day, the Fenland habitat is preserved in Wicken Fen and in a number of shallow " meres " representing still undrained areas of the Peat Fens. Wicken Fen, the property of the National Trust, is carefully guarded and managed by a committee of expert biologists and it illustrates extraordinarily well the difficulty of maintaining nature reserves in a country such as Britain. In Wicken Fen, for example, periodic cutting of reeds is found to be essential to maintain the balance of plant and animal life. It cannot just be left untouched. On the seaward side of the Fen siltlands, the sea is steadily accumulating more silt and as the land becames " ripe " for reclamation it is embanked and reclaimed. This natural process is one which can be encouraged but not unduly hurried : if the whole floor of the Wash were reclaimed by constructing a dyke across its entrance (as has been done with the Zuyder Zee) there is evidence that the land so obtained would be a coarse sandy waste. The fine silt must be allowed first to accumulate.

Mention may be made of one or two outstanding Fenland problems. As the peat is drained it shrinks and so the surface level of the Fen peatlands has been steadily lowered. With exposure to the atmosphere by continued ploughing the organic matter oxidises and disappears and there is growing appreciation that the "structure" of the soil (see p. 96) must be maintained by organic manuring and cannot be maintained by the application of chemical fertilisers only. As the soil loses its

[1] *The Medieval Fenland* and *The Draining of the Fens*, Cambridge University Press, 1940.

crumb structure and as mechanical cultivation in large fields is practised the soil becomes very liable to wind erosion and a new problem is introduced to British agriculture.

A typical Fenland scene is shown in Plate 9a—the land on the right is deliberately flooded to take surplus water which might otherwise cause great damage to the cultivated land on the left.

THE ENGLISH SCARPLANDS

THE ENGLISH Scarplands form a broad belt stretching across England from the Yorkshire coast on the one hand to the coast of Dorset and east Devon on the other and coinciding in general with the outcrop of the Jurassic and Cretaceous rocks. Owing to the regional uplift in the north and west these rocks dip in general towards the south and east except where disturbed by local folding. Across the heart of the Midlands the regional dip is normally to the south-east. The sequence of rocks varies but there is essentially an alternation of " weak " strata with those which are resistant to weathering. According, therefore, to the details of the local sequence and the amount and direction of the dip, one finds developed the varied land forms described on pages 67 to 71. Where the rocks are nearly horizontal one finds a plateau or mesa with fretted edge and outliers; with increasing dip various forms of " cuestas " with the gentle dip slope to south and east and the steep scarp slope facing north, north-west or west. Unlike the northern margin of the Weald or the central fold of the Isle of Wight, the dip is rarely sufficiently steep to give rise to ridges of hog-back type: instead there is an alternation of clay vales of asymmetrical section with ridges coinciding with the outcrop of limestones or sandstones. A map of the scarps has been given above in Fig. 26.

It has already been pointed out that the Jurassic rocks of Britain were laid down in shallow marine waters in a number of distinct basins which were almost separated from one another. Their time of formation was long after the previous great mountain-building period—the Armorican—and there were no great land masses near to yield coarse sediment. The dominant sediments were fine muds and sands but at times at least some of the basins were occupied by waters clear enough to permit the chemical deposition of lime (to form oolites and pisolites) or the growth of corals and the formation of limestones consisting mainly of fragments of shells. Gentle earth movements took place which resulted in the periodic deepening of the basins, but the ridges which separated them remained relatively constant in position. In

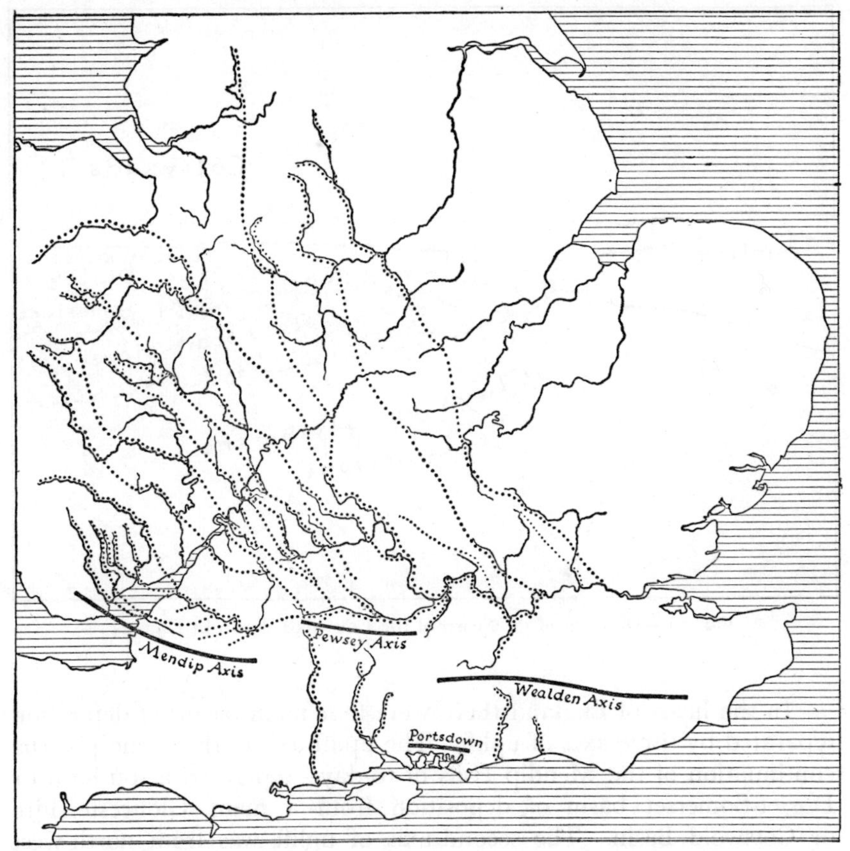

FIG. 69.—Diagram showing the origin of the drainage of the Midlands and south-east of England

Owing to the regional uplift of the west and north, consequent streams developed along the lines shown by dots. Some of the old courses are now deserted, others are occupied by rivers which have been " captured " by other streams (after S. W. Wooldridge).

consequence the sequence of the Jurassic rocks varies widely from one basin to another and this is reflected in the marked changes in the surface features which result. The more massive Jurassic limestones and sandstones give rise to truly noble escarpments (of which the Cotswolds offer a magnificent example) but there is no one Jurassic escarpment stretching across the country. On Fig. 26 are shown the many and varied scarps which exist and their discontinuity is most marked.

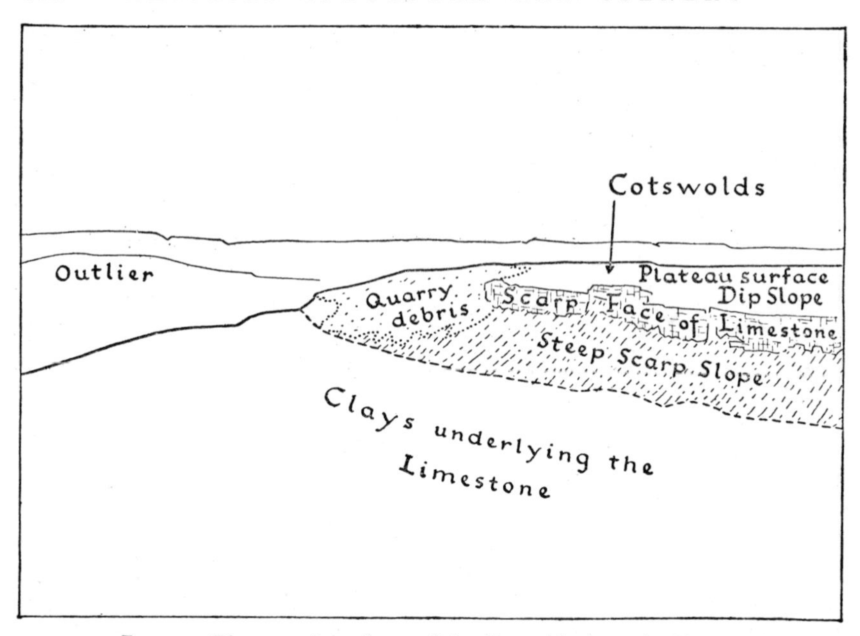

FIG. 70.—Diagram of the Scarp of the Cotswolds shown in Plate 24

In the heart of England there were four main basins of deposition separated by three axes of uplift. The uplift axis of the Mendips—the continuation of the Mendip Hills of to-day—separated a southern or Dorset-Somerset basin of deposition from a main Gloucestershire or Cotswold Basin. The second axis of uplift was from north-west to south-east across where the county of Oxford is now situated. To the north of this lay the great Northamptonshire-Lincolnshire basin. The Market Weighton uplift was across the East Riding of Yorkshire from west to east and to the north lay the North Yorkshire Basin.

No attempt will here be made to describe the very varied sequence of rocks in each of these basins but only to mention a few salient points about each.

In the southern basin, the deposits of which are so well exposed on the Dorset coast, there is a famous sequence of Liassic rocks. Apart from the basal beds where bands of hydraulic limestone alternate with shales as they do practically across England, the sequence is mainly of clays and the important Middle Lias marlstones or ironstones of the Midlands are absent. The Bridport Sands—fine-grained sands yielding a fertile soil—form a conspicuous basal member of the Oolitic sequence : a striking feature is the great thickness of the Oxford Clay and the Kimeridge Clay (separated by the Corallian) and the fine building stone, the Portland Oolite or freestone, near the top of the sequence.

The Liassic beds are very thin over the Mendip axis, but in the Gloucestershire Basin they rapidly thicken and there is a rock bed in the Middle Lias whilst the Upper Lias is sandy. It is in this basin that the Great Oolite is so well developed and, including the Bath Oolite, forms the main Cotswold scarp. The thick Oxford and Kimeridge Clays stretch uninterruptedly across the Mendip Axis and so do the Portland Beds, but the limestone becomes less important. Some of the northernmost occurrences of the Portland limestone are, however, responsible for such well-known features as Shotover Hill, Oxford, where, capped by sandstones of probably Purbeck age, they form the upper part of a conspicuous outlier.

The Oxford axis was important in Liassic and Middle Jurassic times, but later, as with the Mendip axis, its influence was less and the Oxford and Kimeridge Clays sweep across it. Important ironstones, not always on the same horizon, occur in the basin to the north of this axis—the valuable iron ore of the Banbury area is Middle Lias, the Plungar iron ore and the Frodingham ironstone are Lower Lias, the valuable iron ore of south Lincolnshire is Middle Lias. An important bed in the northern part of the basin is the Lincolnshire Limestone, above which is the main ironstone horizon of Northamptonshire. In the northern region the highest Jurassic beds are the Oxford and Kimeridge Clays.

The Market Weighton axis was both important and persistent and the beds deposited in the basin to the north of it are quite different from those in other areas. The Liassic rocks are not dissimilar, but the Middle Jurassic consist mainly of a great series of sandstones—the Lower, Middle and Upper Estuarine Series—and it is these which give rise to the North York moors with the scarps of the Hambledon and Cleveland Hills. These are *not* of limestone.

The Lower Cretaceous Rocks seem to have been deposited in two main basins, a southern and a northern. Throughout the south the predominantly sandy Lower Greensand was succeeded by the Gault.

The latter, a stiff dark clay, has somewhat more sandy beds in the upper part ; these pass towards the west into the sandy " malmstone " and cherts of the Upper Greensand which rest unconformably on older beds to form the well-known flat-topped hills of East Devon (see Plate 5B), such as the Blackdown Hills. In the northern basin the Lower Greensand is represented by the Speeton Clays, with " carstone," succeeded by marls and the famous Red Chalk of Hunstanton and Lincolnshire.

By the time the chalk came to be deposited, tranquil clear water conditions prevailed over much of the country and the effect of the old axes or basins is little seen. Consequently the chalk is continuous as a broad outcrop right across the country, though the scarp form is most marked along the Chilterns and least where it was overridden by the ice-sheets. Even if accidental, the distinction in name between the Chalk " Wolds " of Yorkshire and Lincolnshire and the Chalk "Downs" of Kent, Surrey, Sussex, Hampshire, Dorsetshire and Berkshire actually marks that between the chalklands plastered with drift and the chalklands almost free therefrom. Between the two the Chilterns are not usually referred to as downs and certainly not as wolds, but individual sections of their drift-free scarps and higher sections are called " downs "—e.g. Dunstable Downs. Between the neighbourhood of Royston (Hertfordshire) and the north Norfolk coast the characteristic scarp form does not appear and the chalk outcrop is marked by rolling, open country, largely ploughed but sometimes—as on Newmarket " Heath " and the Gog Magog Hills of Cambridge—in typical fescue pasture.

It is unnecessary to repeat the details given in Chapter 8 of the characteristic land forms of chalk country. With its smooth, rolling outlines, its open branched but waterless valleys, its grey-green fescue grassland, its fields showing white after ploughing where the ploughshare has turned up the chalk underlying the thin soil, it is remarkably distinctive. William Smith, referred to previously as the Father of British Geology, says that he recognised the Yorkshire Wolds as consisting of chalk from a distance of 20 miles before even visiting them.

The age-old practice of " chalking " heavy soils led to the opening up of many small chalk-pits and later many more were needed for the burning of agricultural lime so that chalk country is dotted with quarries. It is obviously easier to quarry into a hill than to dig a pit which probably would have to be drained, so the scarp faces have

The moorland above the Rhondda Valley L. D. STAMP
The long dip-slope of Hirwaun Common (Pennant Grit) in the background
(See page 219 and Fig. 27)

L. D. STAMP
Cwm-parc and Parc Colliery, the Rhondda
The main Rhondda Valley is in the middle distance (See page 219)

M. WIGHT

Near Nevin, Lleyn Peninsula, North Wales
The hills mark outcrops of igneous rocks (See page 217)

M. WIGHT

Gateholm, Pembrokeshire
A typical landscape dominated by a peneplane of marine denudation (See page 221)

become a favourite location for quarries—especially with the development of large-scale industry. Whilst opinions differ, there are many who hold that these dead white slashes in the face of the downs (see Plates 21-22) add variety and interest to the landscape, though a large-scale modern cement plant with its attendant dust (even if smoke and fumes are eliminated) can scarcely be regarded as an attraction. Many of the abandoned chalk-pits become, almost automatically, nature reserves for both plants and animals. The same is true of the sanctuary from the prying hand of man afforded by many railway cuttings in chalk. In old quarries and railway cuttings the chalk face can usually be left with safety almost vertical (see Plate XIII). The white surfaces become black, but the action of frost in winter is constantly flaking off the surface layers and exposing the white chalk once again.

As the Scarplands are traced across England, though the dominant strike is south-west to north-east, there are many local changes in its direction. Some interesting special features are developed where such a change in strike occurs. Thus the change from N.N.W.-S.S.E. in Lincolnshire to N.-S. or N.N.E.-S.S.W. further south has resulted in the great widening of the normal Oxford-Kimeridge Clay Vale and its invasion by waters of the Wash and subsequently by Fenland. A particularly interesting area is in the north-eastern part of Yorkshire. There the rocks dip gently to the south : the Jurassic sandstones build up the North York Moors and pass southwards under a group of clays (partly of Jurassic, partly of Cretaceous, age) which naturally form a broad clay vale before the Chalk Scarp is reached on the south. This clay vale drained originally to the North Sea until the escape of the waters was blocked by ice during the Great Ice Age. The waters were ponded back to form Lake Pickering, from which they eventually spilled over and cut a deep gorge to the south-west. The river has never deserted this course : to-day the Derwent and its tributaries rise a few hundred yards from the North Sea but flow sluggishly inland to escape eventually into the Yorkshire Ouse and so into the Humber. The glacial lake has never quite dried up, for the floor of the Vale of Pickering in places is marshy or peaty. At the western end of the Vale there is an abrupt change of strike : the beds swing round so that the Hambledon Hills, following the strike, take on a north-south direction and impose a western wall to the Vale. It is through this that the Derwent Gorge has been cut (see Fig. 61). Another feature of interest in the Vale is the way in which several parallel streams draining down the dip slope from the North York Moors have cut down in deep narrow valleys to older rocks. On the geological map the outcrops of these older rocks appear as lozenges and both the geological map and the country itself invite a comparison with the " slashed sleeves " of our Elizabethan ancestors.

THE ENGLISH MIDLANDS

THOUGH the country is undulating rather than flat, the heart of England is essentially a plain. With a constant repetition of a rural pattern of small irregular fields enclosed by hedges and in which grassland usually predominates over arable, with scattered farms and villages set in a framework of scattered hedgerow trees (elm and oak predominating) and occasional small woodlands, it is voted by some rather dull : to others it epitomises the quiet rather than the spectacular English countryside.

In the sense in which we shall use the term here, the Midlands are roughly coterminous with the largely drift-covered Triassic plain. The red of the Bunter sands—affording the tracts of lighter soil—as well as of the Keuper marls imparts a dominant red tinge to the soils notwithstanding the intervening drift deposits. On the west the limit of the plain is well defined by the north-south edge of the Welsh massif provided by the Malvern and Abberley Hills, Wyre Forest and the Wrekin : in the north the plain wraps round the southern end of the Pennines. In the south-east the Keuper marls dip under the thin representatives of the Rhaetic, which in turn pass under the Lias. There are sometimes very small scarps to mark the Rhaetic and Lower Lias limestones but the break between the plains developed on the Keuper marls and those on the Lower Lias clays is scarcely perceptible and often the first conspicuous scarp is that of the Middle Lias marlstones. South-westwards there is no break between the Midland plain and the low ground of Gloucestershire—the Vale of Gloucester and Vale of Berkeley through which the Severn flows between the Cotswolds and the Forest of Dean. In the north-west the Midlands are continued through the Midland Gap or Gate into the very similar plain which covers north Shropshire and Cheshire and stretches over the Mersey into south-west Lancashire. This Lancastrian plain to the west of the Pennines is matched by one east of the Pennines extending over Nottinghamshire into the Vale of York.

Many of the main features of the Midlands are surprisingly old, even in geological time. There has been little folding since the waters of the Keuper salt lake lapped round the projecting islands of old rocks. When we climb the Wrekin to-day, or one of the ridges of Charnwood Forest, or even the little Lickey Hills or the Nuneaton ridge, and notice the shattered pre-Cambrian rocks peeping through the surface soil, we can almost visualise the fierce sun of those far-away Triassic times beating down upon the bare rocks and realise that we are looking upon a " fossil landscape." In some ways these " islands " of old rocks—some of them are true *horsts*, or blocks of country surrounded by faults—are one of the most interesting features of the Midlands ; each has a not inconsiderable economic significance and each has some particular features of its own.

Taking these ancient islands in turn : Charnwood Forest consists of a series of ridges built up of pre-Cambrian quartzites and other hard rocks and with some large masses of intrusive igneous rocks—the " granite " of Mount Sorrel, the darker Markfieldite of the Markfield quarries and the microgranite of Groby, all of which provide excellent road metal in a part of the country which requires much and produces little.

To the west the small Leicestershire coalfield is but little exposed at the surface owing to the drift cover and, in contrast to most British coalfields, has given rise to so little industrialisation that collieries in a rural setting are the predominant note.

The long narrow Nuneaton Ridge has also road-metal quarries, but here the associated coalfield has a southern " hidden " extension which has helped, though it was not responsible for, the growth of Coventry.

The Lickey Hills, largely of Cambrian quartzite, having remained uncultivated because of poor soil, provide Birmingham with a playground. Nearby the large and important South Staffordshire Coalfield is directly responsible for that unworthy memorial of the Industrial Revolution, the Black Country. Its famous Thick Coal largely worked out, its local iron ores exhausted, its industrial towns forming now an almost continuous urban agglomeration, the Black Country has survived because of the adaptability of its industries and people. The old tip-heaps are becoming grassed over and it is to be hoped that the time is not far distant when this man-made landscape may again become an area of real beauty. The coalfield has a core of old rocks and already the old Silurian Limestone quarries of the Wren's Nest at Dudley have become an attractive public park. The outcrops of Carboniferous basalt at Rowley Regis are quarried extensively for

Plate XXV

CHICHESTER HARBOUR, SUSSEX

AERO-PICTORIAL

Showing four stages in the silting up of an estuary—the "mud-islands"; the mud-islands coalescing to form a sheet of mud; the mud with drainage channels; the mud enclosed and forming a flat meadow

Plate XXVI

LULWORTH COVE, DORSET

AERO-PICTORIAL

A famous example of differential marine erosion. The outer rampart of cliffs formed by relatively hard Jurassic limestones, when once broken through, exposes to sea erosion the relatively soft clays which have been eaten out into a cicular bay, bounded (on the left) by the resistant chalk

road metal. Further north the coalfield passes under a covering of Bunter sands which have given rise to the sandy heath and forest land, now becoming dotted with collieries, of Cannock Chase.

The other areas of older rock lie on the margins of the Midlands and the Welsh massif. The striking isolated hill of the Wrekin consists mainly of pre-Cambrian rhyolitic lava, again much shattered by eons of weathering, but associated with other rocks including a thin representative of the Carboniferous Limestone. This hill stands at the northern end of the long narrow outcrop of coal measures—the Coalbrookdale and Forest of Wyre field—and it is but a short distance to the south that the Severn cuts through the picturesque Ironbridge Gorge.

Turning now to the outcrops of younger rocks which surround the old islands, they form for the most part lower ground, but some attractive hills may mark the outcrops of the Bunter Sands and Pebble Beds or of the Keuper Waterstones. These beds may, according to their texture, give rise either to light easily worked soils, or to soils so coarse and light as to be agriculturally hungry and so occupied by heathland. Good examples of the latter are Sherwood Forest or the Dukeries—so called because of the large ducal estates laid out on these poor lands—and Cannock Chase. Bunter or Keuper Sandstone scarp features are to be seen in the Helsby and other hills of Cheshire and the Hodnet Hills of Shropshire, whilst similar sandstones provide the low hill which forms the heart of Liverpool as well as the stone for much local building.

Birmingham itself is an almost unique example of a large town situated nearly on a water parting—it was originally sited on relatively poor land as a meeting ground for farmers from the surrounding richer villages.

A proportion of the intervening low ground is developed directly on Keuper marls whilst, especially in Nottinghamshire, the fine-grained sandstones known as the Keuper Waterstones are important as a source of water as well as yielding some good arable soils. But a far larger proportion of the low ground is masked with glacial drift. The old glacial Lake Lapworth (see p. 167) occupied a large area of north Shropshire and Lake Humber stretched over large parts of the Trent Valley as far inland as Nottingham. There are large level areas, passing towards the centre into some ill-drained swampy " moors " or mosses, such as Hatfield Moors and Thorne Moors east of the Pennines or the

Boggy and Weald Moors in central Shropshire. Some of the lake deposits are of the nature of outwash fans and vary greatly in coarseness ; but it is usually beyond the margins of the lakes that are found the irregular masses of glacial sand or the drumlins of boulder clay with " meres " in the hollows. This is well seen in Cheshire ; in the Midlands proper a variable drift is spread unevenly over the surface and is partly responsible for its monotonous character.

On the eastern edge of the Pennines glacial Lake Humber extended as far north as the site of York. York itself affords a classical example of a terminal moraine affording a dry routeway, in this case from east to west, across low-lying swampy ground. Where this east-west land route was crossed by the north-south water route of the Ouse, the Romans established their fort of Eboracum, now the City of York

THE SOUTH-WEST

THE south-western peninsula, coincident in the main with the counties of Devon and Cornwall, is unique amongst the regions of Britain in that though part of Highland Britain and built up of old resistant rocks, highly folded by the Armorican earth movements, it was never covered by the ice-sheets of the great Ice Age and so owes its surface features to sculpturing by other agents. The broad impression of the whole peninsula left in the mind of the visitor is of rolling plateau surfaces, swelling to greater heights in the wind-swept moorlands of Dartmoor, Exmoor, Bodmin Moor and some other moors of lesser extent (see Plate 25). Both with increasing elevation and with increasing exposure to the predominant south-westerly winds, as one passes from Devon into Cornwall trees become scarce and the adjectives " bleak " and " inhospitable " may well be applied to the landscape. But where these rather dull rolling uplands reach the sea in magnificent cliffs, sometimes rising sheer for several hundred feet from clear but ever moving waters, as they do round so much of the long coastline, the interior is forgotten in the glory of sea, cliff and sandy bay.

While the broad impression is thus that of a rolling plateau, seen in detail this is found to be interrupted by numerous deep valleys, often steep-sided and well-wooded and affording shelter which has encouraged the siting of cottages, farm houses and villages as well as affording routeways followed by the picturesque Devonshire lanes. With the warm, wet climate of the south-west these valleys are particularly humid and are well known for their wealth of ferns, whilst the lightness or rarity of frosts permits the survival of plants killed by winter frosts elsewhere in Britain. These sheltered valleys are particularly numerous and large to the south and may there be invaded by arms of the sea to afford such famous estuaries as those of the Helford River, the Fal, Fowey and the Tamar, Dart and Teign rivers (see pages 59-61 and Plate 26).

These present-day features of the south-western peninsula reflect faithfully both the geological structure and the geological history of the area. A glance at the geological map shows that the various rock groups are arranged in a series of broad bands trending from west to east and in this one may recognise at once the Armorican folding of Permo-Carboniferous times. Indeed, the picture which one should get in one's mind's eye is of crustal movements of the utmost intensity folding the whole series of rocks in against the resistant block of Wales to the north, movements so intense that the muds and shales were hardened, minutely puckered and contorted, torn apart, thrust over one another in series of intimate imbricate faults and sometimes imbued even with slatey cleavage. The harder beds, mainly sandstones, were thrown into such intense folds that a structure resembling accordion-pleating results. A single band may be found repeated many times in a single cliff section, often standing on end as vertical strata (Plates 4B, 9B and III) or even overfolded. Although they were so intensely folded the principal effect of the mountain-building movements must have been to erect one great mountain chain in the north and another one in the south, with the younger rocks—of Coal Measure age but without coal seams—preserved in a great central valley or trough. Into the heart of the great southern chain there were injected huge masses of granite which baked the surrounding rocks and seamed them with quartz veins, sometimes bearing ores of tin, copper and radium, and developing in fact all the well-known features of a metamorphic aureole (see page 78).

The rocks which were involved in the Armorican folding were mainly Devonian and Carboniferous, but in the south there are highly altered representatives of Ordovician and Silurian ages whilst pre-Cambrian rocks occur in the Lizard and Start Point and Bolt-head areas. The Devonian rocks are particularly interesting because they were laid down in a sea to the south of the great land masses in the basins of which the Old Red Sandstone rocks were being formed elsewhere in Britain. The north of Devonshire was relatively near that land mass and there the Devonian rocks are mainly shallow-water deposits of sandstones and shales. In south Devonshire the sea water was at intervals sufficiently clear to permit the formation of limestones, which are thus seen in the Devonian sequence and have such a powerful effect on scenery near Torquay and at Plymouth—suggesting comparisons with Mediterranean Riviera resorts with their limestone cliffs and bluffs. Where the Devonian shales have been intensely folded cleavage is sometimes developed ; thus the Delabole

A TOR ON DARTMOOR MINISTRY OF AGRICULTURE

A tor consists of granite, weathered sub-aerially into boulders, standing up above the surrounding surface which probably owes its smooth surface to submarine peneplanations

Plate XXVIII

ROCKY VALLEY, TINTAGEL H.M. GEOLOGICAL SURVEY

Where the old elevated Pliocene platform reaches the sea, surface streams drop almost sheer to the ocean. Notice the large pothole

slates from the famous quarries near the village of that name are true cleaved slates of Devonian age. The Carboniferous rocks which succeed the Devonian are different from those in other parts of England. They comprise a monotonous series of dark shales, with numerous bands of sandstone, resembling the Coal Measures elsewhere but without coal seams and containing scanty plant remains which show that they represent the whole Carboniferous sequence from the Carboniferous Limestone to the Coal Measures.

It may be said that the geological history of the south-western peninsula from those remote times has been essentially the story of the gradual attrition of the Armorican mountains. The great tectonic basin between the northern and southern ranges was like a desert basin of the present day. In a few depressions red clays such as the Watcombe Clays of the Torquay-Teignmouth district were laid down, but the obvious deposit to be formed was of coarse breccias and conglomerates comparable with those being formed in parts of Persia at the present day. These breccias are really fossil screes of angular blocks of rock and fragments—the angularity being attributable to sun and frost action—mixed with boulders and stones rounded by torrents after heavy storms. They are often banked up against the parent rocks and are usually thicker in the west than in the east. They are succeeded by conglomerates and red sandstones—up to 2000 feet thick near Exmouth—which are of the nature of outwash fans. In both the breccias and sandstones bands of marl with sun-cracks and ripple marks point to the existence of temporary lakes or pools during wet periods. The marls which succeed the sandstones are also red, but their more fine grained character suggests that the surrounding mountains had already been considerably worn down. These Red Permian beds are succeeded normally by the Triassic, beginning with the famous Budleigh Salterton Pebble Beds of well-rolled pebbles, probably laid down by torrential rushes of water from the western land mass, and succeeded by red sandstones and marls. It is this varied succession of red beds of Permian and Triassic ages which gives rise to the well-known red soils of East Devon—so characteristic that the area is often called Red Devon. The sandstones give warm, light, easily worked soils, the marls much heavier ones.

Returning to the main area further west, by Triassic times there were scarcely more than stumps left of the Armorican Mountains. The granites of the central core had been exposed and the varying resistance to weathering of the different rock types had exerted an influence which has persisted to the present day. The main mass of Devonian sandstones proved tough and it is these rocks which form the upstanding moorland of Exmoor. The granites are resistant to weathering, consequently the largest granite mass, that of Dartmoor, forms both the most extensive and the highest moorland mountain mass of the south-west. In their turn each of the granite masses—Bodmin, St. Austell, Camborne-Redruth, Carnmenellis, Land's End and Scilly Isles—gives rise to land at differing elevations but consistently above the level of the surrounding country. Another upstanding mass is that of the Lizard Peninsula, coinciding with a complex massif of basic igneous rocks (the famous Lizard serpentine), intermediate rocks (gabbro), some granite and some ancient gneisses. The whole mass is probably pre-Cambrian. There are smaller upland masses which mark the outcrop of the harder or more extensive sandstone horizons—such as the Staddon Grits of Cornwall which form the St. Breock Downs—but in general the complex folds in the various sedimentary rocks seem to have little or no effect on the form of the surface. Why should this be the case? It is scarcely necessary to trace the geographical evolution of the south-west peninsula in detail from Triassic times to the present day: it is sufficient to say that it has been subjected to eustatic or up-and-down movements but little folding and that both when above the level of the sea as well as when below it has been subjected to long-continued peneplanation. It is often difficult to say which of the peneplanes, of which there are substantial remnants at different levels, are due to sub-aerial peneplanation and which to sub-marine. In both cases there are a few outstanding masses or monadnocks of which the "tors" such as Yes Tor, Rough Tor, and Brown Willy, are good examples. There are traces of a high level peneplane at about 2000 feet, but it is from the 1000-foot plane, well seen on the eastern side of Bodmin Moor and the south-eastern side of Dartmoor, that the finest tors are seen to rise—notably the Cheesewring (see also Plate XXVII). There is a middle platform at about 700 to 800 feet, seen for example around Moretonhampstead and also around Camelford. Best marked of all is the 400-foot platform which has been established as Pliocene in age (and so comparable with the 400-foot platform in the London Basin and south-eastern England) and marine in origin since it is frequently bounded inland by an old sea-cliff.

These peneplaned surfaces are of more than academic interest. They provide large stretches of land at several elevations but with little slope so that drainage is slow. This is serious in such a damp climate and as the planes stretch uninterruptedly across sandstone and shales, with a tendency for hollows to form on the shales there is evidence of seriously impeded drainage in many areas. The shales weather to a sticky blue or brown clay and the soils have a mottled gley horizon (see page 94) and support a vegetation in which cotton grass, sphagnum and Molinia are in evidence (see Plate 4A).

Although, as already stated, the south-western peninsula was little affected by earth-movements after the Armorican, the Alpine orogeny did result in the development of faulted belts along which the rocks were crushed and river erosion facilitated. The north-west to south-east course of the Teign east of Dartmoor is explained by these fault lines, whilst the little Tertiary basin of Bovey Tracey with its lignite deposits is a tectonic basin of Tertiary age.

The post-pliocene uplift of the whole area—so that the main Pliocene platform now lies at some 400 feet above sea level—has had two most important effects. The first is the rejuvenation of the rivers so that they occupy deep, gorge-like courses, often in the midst of traces of their former mature valleys at high levels. Some of the gorges are famous, such as Lydford Gorge now the property of the National Trust. The second is the target which has been offered to the attacking seawaves, especially to the great breakers which roll in from 4000 miles of uninterrupted ocean on the north coast of Cornwall. The main direction of the waves is from the west and the "grain" of the country is east-west. Consequently ribs of rock run out into the ocean whilst bays have been excavated along the outcrops of softer rock and lie picturesquely between rocky headlands (see Plate 3B).

The post-Pliocene uplift went on intermittently so that intermediate wave-cut platforms were formed, until the land stood higher than it does at present and the rivers cut down their valleys to below their present levels. Near their mouths forests grew and when later the land sank these forests were submerged so that their remains may be seen, notably around Torbay, at or below low water mark. At the same time the sea invaded the valleys of the rivers and gave rise to those beautiful branching estuaries or rias which are such a feature of the south coast (see Plates 26A and 26B).

The entry to the region we have just described is the cathedral city of Exeter. That part of Devonshire which lies to the east of Exeter belongs to lowland Britain and has been so described, but by way of exchange the Quantocks of Somerset are, both geologically and geographically, a detached portion of Exmoor.

Dartmoor has recently been fully described by L. A. Harvey and D. St. Leger-Gordon in the volume published in the *New Naturalist* series.

THE WELSH MASSIF

THIS chapter is entitled the Welsh Massif rather than Wales because the clearly defined block of Highland Britain which is occupied by the Principality extends beyond its borders to include also the southern half of Shropshire, the whole of Herefordshire and the whole of Monmouthshire. On the other hand the fertile Vale of Glamorgan, south of the coalfield rim, belongs structurally and physically to Lowland Britain.

It will be clear from the earlier chapters of this book that the ancient earth-block which we have called the Welsh Massif was built up by successive orogenic movements but that its surface features are in large measure due to long-continued denudation during geological periods which have otherwise left little trace. If we take first the process of building up we find the largest stretch of pre-Cambrian rocks—all highly folded and intensely altered metamorphic rocks—in the Isle of Anglesey, the ancient Mona, which thus gives its name to the Mona complex. The numerous folds in the island trend south-west to north-east and this is the typical Caledonian trend, but Carboniferous Limestone is involved in some of the folds. We may say that the main folding of Anglesey was pre-Cambrian and Caledonian and that Carboniferous Limestone was deposited on an ancient surface of varied character but that renewed movement or posthumous folding in Armorican times took place along the old lines. Broadly, however, Anglesey is the oldest part of the Welsh massif and it was against this block that later folding took place.

The rugged mountains of North Wales consist essentially of Cambrian, Ordovician and Silurian rocks : the higher mountains of to-day coincide largely with the outcrops of resistant igneous rocks—lavas and sills—especially of Ordovician age. The lavas seem not only to have resisted subsequent denudation but also the more intense folding. The structure of the Snowdon Syncline and the Harlech Dome which succeeds it to the south are relatively simple and both are essentially Caledonian in trend. Southwards into Central Wales volcanic rocks play a minor rôle, but the monotonous shales and mudstones are often

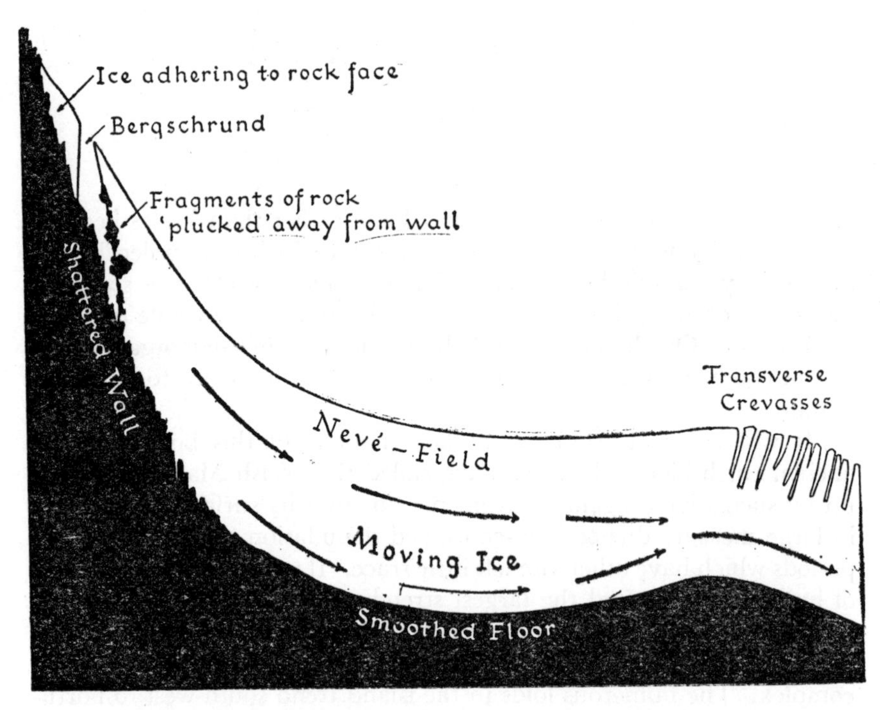

FIG. 71.—Diagram showing the formation of a cirque (after Holmes)

intensely folded and crumpled, thrust faults are common and, generally, the rocks are twisted and torn. Yet because of the absence of great hard beds the scenery of Central Wales, by contrast with North Wales, is relatively tame. But still, in Central Wales, the folding is essentially Caledonian though there was often renewed movement along the old lines in later periods.

South Wales is definitely dominated by the great synclinal fold or basin which is typically Armorican with a dominant east-west strike. This is the basin of the South Wales Coalfield. In geological structure it is shaped very much like a pie-dish. Its " rim " consists of hard beds of sandstone and conglomerate of the upper Old Red Sandstone and the Carboniferous Limestone which succeeds it. These rocks are resistant to weathering and so do actually form an upstanding rim, the average height of which is well over 1000 feet above sea level.

Inside the rim there is a thin representative of the Millstone Grit and then the Coal Measures, which are clearly grouped into two coal-bearing series of shales and sandstones, separated by a thick group of massive sandstones, known as the Pennant Grit. It is this Grit which is responsible for the wide stretches of moorland, much over 1000 feet and actually reaching 1969 feet, which separate the famous narrow valleys so typical of the South Wales Coalfield. These valleys follow two chief directions: N.N.W. to S.S.E. and E.N.E. to W.S.W., and to a large extent have been determined by lines of faulting. Many of the first group especially are what the Americans call "shoe-string" valleys: they have no outlet at the upper end except by a stiff climb over moorland. Until the close of the last war there was no way out of those upper ends for wheeled traffic, but now, partly through the work of prisoners, some excellently graded roads, with the aid of hairpin bends, climb over the intervening divides, though even these roads have done but little to break down the extraordinary isolation of the valleys and their human occupants (see Plates 13 and 27).

The South Wales synclinal coal basin is asymmetrical: along the northern rim the rocks dip gently towards the south but along the southern rim they dip steeply to the north or are almost vertical. In parts of the east there is a kink or upfold in the floor of the basin which fortunately brings the lower coal-bearing series quite close to the surface and enables the coals to be worked at a reasonable depth.

To the west the basin narrows and tails out into Pembrokeshire: to the south the Carboniferous Limestone is gently folded so as to form a considerable expanse in the Gower Peninsula and again in Pembrokeshire. Incidentally in north Pembrokeshire the Ordovician sequence is again varied by different volcanic rocks and some intense folding brings even pre-Cambrian rocks into the picture.

In a previous chapter it has been shown that when the Armorican earth movements folded the rocks up against the pre-existing earth blocks both east-west and north-south folds resulted. Thus the strike of the rocks in the east end of the South Wales Coalfield is north-south and the high ground of the coalfield gives place abruptly to low pastoral country on the marls of the Old Red Sandstone. Further east, where two synclines cross one another, the almost circular coal basin of the Forest of Dean coalfield repeats in miniature very many of the features of the South Wales Coalfield itself.

The Welsh Massif is bounded on the east by a succession of remarkable fold-ranges which trend from north to south and of which the most spectacular is undoubtedly that of the Malvern Hills. This is a much-faulted and geologically very complex fold of anticlinal nature, consisting of a core of pre-Cambrian rocks and various Ordovician and Silurian rocks on the flanks. The magnificent ridge overlooks the Vale of Gloucester and the English Midlands on the east and it is continued at a lower elevation to the north in the Abberley Hills (Plate 17).

Between the ridge of the Malverns on the east, the South Wales and Forest of Dean coalfields on the south and the Silurian outcrops on the north-west is a large, roughly triangular area embracing practically the whole of Herefordshire and parts of neighbouring counties, which the geological map shows to consist also entirely of Old Red Sandstone except where interrupted by upfolds of Silurian. This tract of country varies extraordinarily in character. Where the Lower Old Red Marls occur (especially where they afford soils calcareous owing to the occurrence of cornstones or nodular limestones) they give rise to undulating, fertile farmland—the fine country of the so-called Plain of Hereford. Where, however, the dominant rocks are sandstone may be found some of the wildest and most barren moorlands in all Wales—the Black Mountains and Mynydd Eppynt.

Northwards the Hereford Plain gives place to the fascinating country of south Shropshire—classic ground to the geologist. Here the dominant strike of the rocks is from south-west to north-east and rocks from pre-Cambrian to Coal Measure Age are exposed. A mass of pre-Cambrian sediments gives rise to that upland moorland known as the Longmynd : pre-Cambrian volcanic rocks form the upstanding masses of Caer Caradoc, the Lawley, the Wrekin and Pontesford Hill. Against the pre-Cambrian are the representatives of the Cambrian and some upper Ordovician but the interest of the country centres on the Silurian rocks. These consist of an alternating succession of limestones and shales. The two principal limestones, the Wenlock and Aymestry, give rise to two fine scarps—Wenlock Edge and Axe Edge, the one overlooking the fine fertile valley of Corve Dale and the other the almost equally fine valley of Ape Dale.

To the south-east of the county of Shropshire are those striking hills the Clees, which owe their existence to a capping of basalt (much quarried as road metal) of Carboniferous age. These do not by any means exhaust the features of that part of Shropshire which lies within the Welsh Massif: the gorge of the Severn through Coalbrookdale and

Plate XXIX

LLYN-Y-CAE, CADER IDRIS, NORTH WALES

A magnificent example of a corrie lake, with frost-shattered rocks in the foreground

F. FRITH AND CO.

Plate XXX

GIGGLESWICK SCAR

A. HORNER AND SONS

One of the finest examples of a fault scarp in Britain. Carboniferous Limestone on the left, downfaulted shales of the Millstone Grit on the right

Ironbridge ; the synclinal mass capped by Old Red Sandstone of Long Mountain ; the dolerite hills of the Breiddens ; the famous dolerite laccolite of the Corndon all call for notice.

Further north, between the Silurian shales and sandstones of North Wales and the Cheshire plain, are outcrops of Carboniferous Limestone resting on the old rocks and dipping eastwards, giving place on lower ground to the Coal Measures of the North Wales fields. In Denbigh and Flint the sequence may be repeated by strike faulting and the Vale of Clwyd is one of the results.

So much for the general picture presented by the " solid " geology of Wales. Reasons have been given above for believing that the peneplanation of the mass took place in Triassic and Jurassic times and was perhaps finished by the Cretaceous Sea and that the " concordant summit levels " of the higher Welsh mountains represent all that is left of this peneplane. It is instructive to take a physical map of Wales and notice in how many cases the rivers have N.W. to S.E. sections to their courses which suggests consequent drainage of the uplifted peneplane. These rivers must have supplied much material throughout Tertiary times and hence Wales has been subjected to sub-aerial denudation for many millions of years and was already sculptured into much its present form before the glaciers and ice-caps of the Great Ice Age put the finishing touches. North Wales is, of course, classic ground for the observance of many features of glacial origin—the cirques or cwms with their lakes (Plates 8A, XXIX, and Fig. 36), the pyramidal peak of Snowdon with its radiating arêtes, the roches moutonnées of Llanberis, these are all text book examples. So also are the U-shaped valleys of the Pass of Llanberis (Plate 18) and of Tal-y-llyn to the south of Cader Idris, whilst on lower ground the southward diversion of the upper Severn is a magnificent example of glacial interference with drainage.

As with the South-Western Peninsula, South Wales (notably the Gower Peninsula and Pembrokeshire) exhibits magnificent examples of sub-marine peneplanation with platforms at different levels and raised beaches, several of which are illustrated in the coloured plates (Plates 5A and 28B, see also Plate XXB).

THE NORTH OF ENGLAND— THE LAKES AND THE PENNINES

OF THE several sections which go to make up Highland Britain, few surpass in interest England's Lake District. The attraction which it has for the open air holidaymaker is at least matched in its fascination for the geologist. Many of its secrets are still firmly held and the most diverse views have been expressed regarding its structure, but the interpretation which follows is that now generally accepted.

The core of the Lake District is a short section of one of the Caledonides and its rocks have the typical Caledonian trend from south-west to north-east. Though puckered by many minor folds, the structure is essentially anticlinal. The oldest rocks, in the centre of the anticline, are a monotonous series of dark slates, the Skiddaw Slates or Skiddavian. They are partly upper Cambrian and partly lower Ordovician in age and, though resistant to weathering, the mountains such as Skiddaw which consist of them have rounded rather than craggy outlines. The deposition of the Skiddaw Slates was followed by the deposition of a vast thickness of volcanic rocks of various types, but gave rise to craggy mountains of which Helvellyn and Langdale Pikes are good examples (see Plate XVIb). They outcrop now over a narrow belt *north* of the Skiddaw Slates and over a very broad belt—the real heart of the Lakes—*south* of them. It is clear that after the deposition of the Volcanics the area was subjected to a pre-Caledonian folding, for the succeeding Coniston Limestone rocks (Bala age) rest unconformably on the older rocks and are much less severely folded. It would seem that, after this, shallow water resulted for the Silurian strata—Llandovery, Wenlock and Ludlow—form a monotonous series of flagstones or shales. These also give rise to the rather tamer scenery associated with the southern part of the Lake District.

After the main Caledonian folding it is probable that the Lake District range divided the Caledonian Old Red Basin from the separate basin or gulf to the south. When much later the worn-down remnants sank beneath the gentle waves of the Carboniferous Limestone sea it is probable that the limestone was deposited over the whole, but of this we cannot be certain. The limestone is followed by a thin representative of the Millstone Grit, then by the Coal Measures, the New Red Sandstone and the Jurassic. How far all, or any, of these were deposited over the heart of the Lakes is unknown. What is quite certain is that at some later stage, probably through underground movement of masses of magma, the *centre* of the area was uplifted. A roughly radial drainage system developed, probably not unconnected with radial cracking of the now rigid block, and the Carboniferous rocks were weathered to form an irregular succession of inward-facing scarps. Part of the uplift may have been Armorican—connected with the west-east Market Weighton axis—but in the main it is believed to be Alpine. No mention has been made of the many large masses of granite which complicate both the structure and the relief. The age of these masses is uncertain—some may be Armorican, others are believed to be Tertiary and may be associated with the uplift which caused the radial drainage. Because of the concentric rings of Carboniferous Limestone and later rocks and the radial drainage, geologists often refer to the " Cumbrian Dome " but in the form of the ground there is nothing to suggest a dome as there is with the Harlech Dome.

The Carboniferous Limestone may be seen at intervals especially to the north and to the south : on higher ground rock outcrops are numerous and the attractive way in which it weathers make weathered " Westmorland Limestone " one of the most popular materials for suburban rock gardens as far afield as London. On lower ground the limestone is well wooded. The main tract of Coal Measures lies to the north-west—the West Cumberland Coalfield—but the surface outcrops are largely obscured by glacial deposits. In the later stage of the Ice Age, the Lake District had an ice-cap of its own : and the radiating tongues of ice widened and deepened the pre-existing valleys, smoothing their sides and removing the spurs. When the ice finally retreated moraines blocked the mouths of many of the valleys and to-day these help to hold up the lakes which are the distinctive features of the area. Waterfalls from hanging valleys (p. 87) add to the scenic beauties. At the same time, the material swept from the central heights was scattered over the surrounding lowlands of the Eden Valley, the Solway Plain, the west coast and the head of Morecambe Bay, in some of those tracts to be mingled with material brought by other glaciers from Scotland and the Irish Sea. The detailed physiography of the Solway Firth lowlands is that of a recent glacial landscape, with well-preserved drumlins and moraines. The harder beds of the underlying rocks, such as the Penrith Sandstone, may outcrop at the surface from under the mantle of glacial deposits. The St. Bees Sandstone, also of New Red Sandstone age, forms the red cliffs of St. Bees Head (see Plate 12): to the south are stretches of sand dunes whilst along Solway Firth are salt marshes and raised beach terraces in remarkable variety, thus completing one of the most fascinatingly varied of all British regions.

The Isle of Man has structurally many points in common with the Lake District. The Manx Slates which build up the main mass of moorland of the island, including Snaefell, are probably the same as the Skiddaw Slates and doubtless formed part of the same Caledonian range. Carboniferous Limestone rests unconformably on the Manx Slates as it does on the older Palaeozoics of the Lakes, and is conspicuous in the south. The flat northern end of the island affords an interesting series of raised beaches.

The Pennines or Pennine Upland are often wrongly referred to as a "chain" but scarcely form even a "range." They are relatively simple in structure and relatively homogeneous in their rock constituents. Essentially the structure is that of a north-south anticline, of Armorican age, built up of Carboniferous Limestone and Millstone Grit, together with the more local developments of shales—the Yoredale and Pendleside series—which lie between them. In the north the anticline is asymmetric with a very steep and faulted western limit—hence the 2000 foot high scarp of the Cross Fell mass overlooking the Eden Valley—and a gentle eastern limb. Over large areas the Millstone Grit forms the surface rocks and the Carboniferous Limestone is not exposed at all or only in deep valleys. In this northern area the structure may be complicated in three ways: by giant cross faults such as those known as the Dent and Craven faults by Settle with the great limestone "scars" of Ingleborough (Plate XXX); by intrusive masses of dolerite or basalt, especially the famous Whin Sill and attendant dykes; and by minor folding. South of the High Peak which, despite its name, is a flat-topped mass of Millstone Grit, the anticline becomes more symmetrical with Carboniferous Limestone exposed in the centre—well seen in the gorge-like valleys of the Manifold, Dove, Derwent and other streams—flanked by inward-facing

Plate XXXI

H.M. GEOLOGICAL SURVEY

SUCCESSIVE ALLUVIAL TERRACES OF THE RIVER FINDHORN, SCOTLAND

These terraces are cut out of glacial deposits and mark stages in the lowering of the river bed. Looking downstream from near Daless, Nairn

Plate XXXII

STAC LEE, ST. KILDA R.A.F.

The St. Kilda group of islands consists of a remarkable igneous complex of which this steep-sided rock is a part

scarps of Millstone Grit. On the western margins lies the much faulted coalfield of North Staffordshire ; on the east the great coalfield of Yorkshire, Nottinghamshire and Derbyshire. On this eastern side the outcrops of the main sandstone bands in the Coal Measure give rise to other west-facing scarps, thus continuing in series those formed by the Millstone Grit.

Some special features of interest in the Pennines call for comment. One is the importance of the gaps by which they can be crossed with relative ease from east to west. In the north is the Hexham, Haltwhistle or Tyne Gap along the northern side of which stands that monument of Roman times, Hadrian's Wall, stretching from Newcastle to Carlisle. The second gap is that of Stainmoor, just to the south of which the Shap Fells link the Pennines with the Lakes and offer a barrier 1000 feet high to be surmounted by the west coast railway route to Scotland. The third gap is the Aire Gap—up the valley of the Aire from Leeds and across the divide to the valley of the Ribble. Elsewhere the Pennines remain a formidable barrier to be crossed by tunnel or by roads which may be blocked by snow for several days, if not longer, each winter.

North of the Aire Gap the Pennines give out a western spur into Lancashire—the Bowland Fells or Bowland Forest. South of the gap is a similar spur, the Rossendale Fells, around which lie the Lancastrian Coalfield and the cotton towns. Such names as Ashton-under-Lyne or, further south, Newcastle-under-Lyme, suggest their situation under the steep rise to the Pennines.

Though the Pennines form on the whole a " negative area " (mainly uncultivated and almost uninhabited) they have so many functions that the problem is to integrate their multiple use—as gathering grounds for water supply, as sheep pastures, as grouse moors, as recreational areas for the neighbouring great towns. The Millstone Grit areas are the chief areas involved because of the fine soft waters and heather on the better drained parts : the Limestone Pennines are less important for supply of water though affording excellent fescue pastures and spectacular rock scenery.

Northumberland and Durham are structurally the eastern flanks of the northern parts of the Pennines. In Northumberland the surface level drops steadily to the east till interrupted by the fine scarp made by the Fell Sandstones (see p. 116) ; in Durham the main interruption is afforded by the Magnesian Limestone scarp. All the lower ground in

Northumberland and Durham is plastered with glacial drift and the general character of the eastern parts of the two counties is that of a low plateau, drift-covered, through which the streams have cut deep "denes"—steep-sided, wooded, gorge-like valleys. The valley of the Wear at Durham is, of course, the supreme example of this.

SCOTLAND

THE EXISTENCE of a separate volume in this series devoted exclusively to the Highlands of Scotland renders permissible the otherwise impossible attempt to deal with the whole of the structure and scenery of Scotland in a single short chapter.

Scotland falls very simply into three divisions—the Highlands and Islands, the Southern Uplands and, lying between those two mainly hilly areas, the Central Lowlands. On both sides the Central Lowlands are bounded by series of faults and the direction of these, from north-east to south-west, strikes the dominant note in the structure of the whole of Scotland—the determination of its main features by the Siluro-Devonian or Caledonian earth-movements. In essence, the Highlands and the Southern Uplands are the worn-down remnants of the Caledonides ; between them is the rift valley initiated at the same time but which has been subjected both to renewed movement (posthumous " Caledonian " folding mainly of Armorican date) as well as the separation of the once continuous spread of Carboniferous deposits of its floor into separate basins by east-west and north-south folds of Armorican date.

The region known to geographers as the Southern Uplands coincides with what is commonly regarded as Southern Scotland—an almost continuous belt of high ground stretching from the Irish Channel to the North Sea and bounded on the north by the line of faults which form the southern boundary of the Central Lowlands or Midland Valley. Along this northern line the Lammermuir Hills rise sharply to heights of 1500 to 1700 feet above the plain of East Lothian and a little to the west the Moorfoot Hills overlooking the Midlothian coal basin rise to even greater heights. Westwards the Uplands proper fade rather more gradually into the higher ground of the Central Lowlands though a fine scarp edge overlooks part of the valley of the upper Clyde where that river flows from south-west to north-east, and near by the Lowthers or Lead Hills rise to 2408 feet.

Over the Uplands themselves there are two distinct types of scenery. From Nithsdale almost to the Berwickshire coast the hills are rounded, smooth and grass-covered—with ill-drained summit mosses in some cases and heather-covered sides. Over a large central area they reach a common height (Plate XXIIB) which is an elevated peneplane traversed by deep and narrow valleys such as that of the Moffat Water. This peneplaned surface is probably pre-Jurassic or at least pre-Cretaceous and the course of the rivers suggests their initiation on a south-eastward sloping surface of Cretaceous or later sediments of which all traces have now disappeared. Eastwards, where the Uplands give place to the fertile valley of the Tweed, the scenery is diversified by picturesque isolated hills which are either old volcanic necks or, as in the case of the Eildon Hills near Melrose, the actual remains of old groups of volcanoes complete with lava flows and intrusions. To the south here rise the Cheviot Hills formed of great masses of volcanic rock of Old Red Sandstone age. South-westwards of Nithsdale the scenery is different owing to the existence of large masses of granite—notably the three great masses of Criffel or Dalbeattie, Cairnsmore of Fleet and Loch Doon, as well as the smaller masses of Cairnsmore of Carsphairn, Spango and The Knipe—which offer scenes of wild and rugged grandeur and many points exceeding 2000 feet. Amongst the highly folded Ordovician and Silurian rocks—there are none older unless certain rocks believed by some geologists to be contemporary intrusives are in reality the old metamorphic floor on which the Ordovician rocks were deposited—are the basins occupied by the Old Red Sandstone, notably the Tweed basin with lower Teviotdale. The Old Red Sandstone was followed by the deposition of Carboniferous rocks. In places these are coal-bearing, with the result that small coal basins, of which the chief is that of Sanquhar, are found rather unexpectedly. Along the Irish Sea coasts are lowlands of varied origin—those of Dumfries, the lower Annan and the Stranraer Isthmus coinciding with New Red Sandstone basins—though everywhere on the low ground there is a masking of the old rocks by boulder clay.

The terms Central Lowland and Midland Valley are used interchangeably for the populous heart of Scotland. Both are inapt; the area is structurally a rift valley let down between parallel faults but is not otherwise a valley in the ordinarily accepted meaning of the word: the area is only a lowland by comparison with the great masses of the Highlands on the one side and the Southern Uplands on the other. Much of its surface is several hundred feet above sea level and it includes such hill-masses as Campsie Fells and the Ochils, of which

L. D. STAMP

Snowdon from the east, with Llynau Mymbyr in the foreground (See Fig. 29, and pages 70 and 217)

J. A. JENSON

The Scafell range from the summit of Great Gable (See page 222)

J. A. JENSON

The Screes, Wastwater (See page 21 and page 46)

some points rise to over 2000 feet. It does, however, include a large proportion of the lower land of Scotland and is the home of more than four-fifths of Scottish people. Initiated in Caledonian times, the Rift Valley became a basin of deposition of Old Red Sandstone rocks, accompanied by much volcanic activity so that contemporary lava flows, volcanic necks, sills and other intrusions are numerous and widespread and by their resistance to weathering are the cause of many of the upstanding hill-masses. The deposition of Old Red rocks was followed by swamp conditions resulting in the growth of forests. Similar conditions recurred in the later parts of the Carboniferous period so that coal seams occur in both earlier and later Carboniferous rocks. The Armorican crustal movements accentuated the old fold and fault lines but at the same time created folds with east-west and north-south axes which resulted in the present day separation of the coal-bearing beds into basins through denudation along the anticlinal ridges. The folding continued for a long period for it affects the Permian Mauchline Sandstones. These extremely interesting bright brick-red or orange-red sandstones occur in Ayrshire and are believed to represent desert sand-dunes for they exhibit dune-bedding on a huge scale. Both the Carboniferous and Permian periods witnessed widespread volcanic activity and the geology of the Midland Valley is still further complicated by the presence of dykes of Tertiary age. Apart from the latter, however, there are no rocks to represent the enormous space of time between the Permian and the Great Ice Age. Many geologists believe the whole Midland Valley to have been covered by Mesozoic and perhaps Tertiary sediments and that the river systems, with the marked north-west to south-east courses of so many rivers, were initiated by consequent streams flowing off the Highlands over a plain of sediments tilted south-eastwards and stretching right across the Southern Uplands.

Turning to the present-day scenery of the Midland Valley, the Highland Boundary Faults give rise to one of the finest examples of a fault-scarp to be found anywhere—the great wall of the Highlands overlooking a valley excavated in Old Red Sandstone rocks. This valley can be traced from the Clyde to near Stonehaven and between Methven and Perth it broadens out to form the undulating plain known as Strathmore. South of this is a long belt of high ground—the Garvock, Sidlaw (1492 feet) and Ochil (2363 feet) Hills built up of volcanic rocks of Lower Old Red Sandstone age and the Campsie Fells

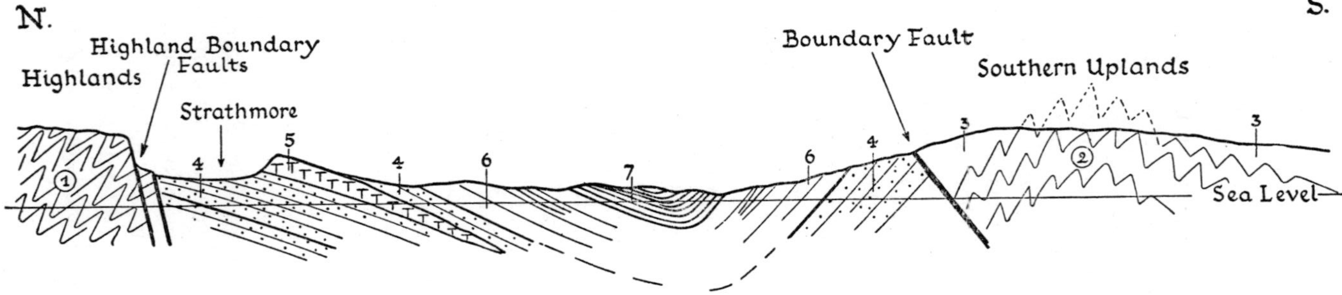

Fig. 72.—Section through the Rift Valley of Scotland

(1894 feet) and Kilpatrick Hills (1313 feet) of Lower Carboniferous volcanics. The line is continued by similar rocks on the south side of the Clyde in a broad mass of upland culminating in the Hill of Stake (1711 feet). As a result of the vertical jointing commonly developed in volcanic rocks, many of these masses have precipitous sides and relatively flat tops. The gap in the line between the Ochils and Sidlaws is formed by the Tay and this natural gateway to the Highlands is guarded by Perth, just as Stirling commands the gap formed by the Forth between the Ochils and the Campsie Fells.

Along the southern margin of the Midland Valley there is a much more interrupted belt of hilly ground fringing, or forming offshoots from, the Southern Uplands. On the eastern side of the Midlothian Coalfield plain are the volcanic Garleton Hills, to the west the long anticlinal ridge of the Pentland Hills with its core of highly folded Silurian rocks and its fringing masses of Old Red Sandstone volcanics and sediments, including coarse conglomerates which are usually resistant to weathering. Similar conglomerates for the same reason build up Cairntable (1944 feet) whilst amongst other masses formed by igneous rocks Tinto Hill (2335 feet) may be mentioned as an interesting example of a laccolite (see p. 79).

A broad upland of lower Carboniferous lavas separates the two main lowland areas—the Ayrshire basin and the main central tract stretching across the narrow waist of Scotland from the lower Clyde in the neighbourhood of Paisley to the Forth and extending into the Fife Peninsula. Even this area is very far from being a featureless plain. There are many hills due to intrusive sills (such as the dolerite hills of Fife) or bosses and volcanic necks (such as Stirling Castle Rock) or to outpourings of lava. Most of the lower ground and not a little of the higher is plastered with glacial drift which results in many areas of poor natural drainage, whilst the fascinating crag-and-tail structure associated with many of the isolated hills has already been described (see p. 85). Glacial lake basins, outwash fan areas and raised beaches (the latter particularly noteworthy along the lower Clyde) add to the diversity.

The Geological Survey has adopted the name Grampian Highlands for that part of Scotland which lies between the two great fault-lines of the Great Glen and the Highland Boundary. It is unnecessary here to enter into the many still unsolved problems of the structure of the Grampians. It is difficult for the uninitiated to believe that the quietly

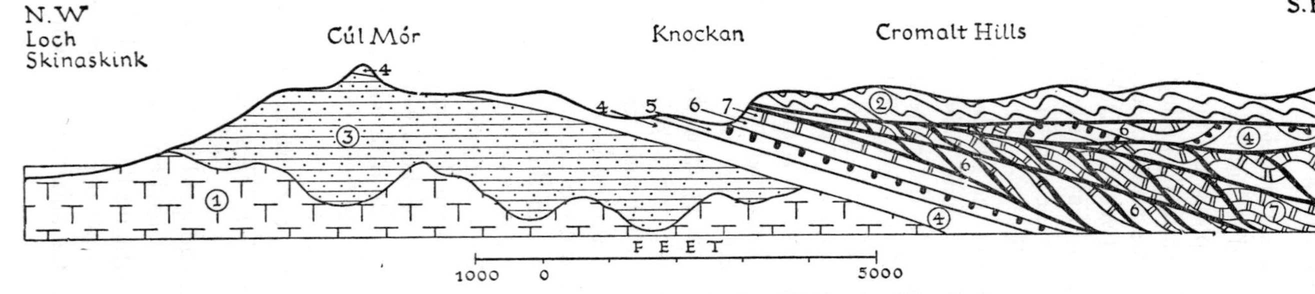

FIG. 73.—Section through the North-West Highlands of Scotland

1. The ancient crystalline complex (Lewisian).
2. Other ancient crystalline rocks, similar to those found in the Grampians, which have been thrust westwards over younger strata.
3. Torridonian Sandstones; 4-7 Cambrian rocks including the Durness Limestone (7).
Heavy lines represent thrust faults or planes over which masses of rock have been thrust westwards, sometimes in the complex "imbricate" structure.

rounded outline of the Cairngorms may hide structures as complex as any to be found in the wildest parts of the Alps or Himalayas, yet such is the case. We are, in fact, dealing with the worn-down stumps of the greatest of the Caledonides. Peneplanation has been going on intermittently since Old Red Sandstone times and much can be removed in the course of what may be some 200,000,000 years. To-day the heart of the Grampians or Central Highlands consists of a high table-land—in fact an elevated peneplane—with a general level of 2000 to 3000 feet above sea-level. The eastern part of this is relatively little dissected and has a few corrie-lochs only. Though a considerable area in the Cairngorms rises to over 4000 feet, these higher lands form gentle swellings rising from the plateau surface rather than rugged peaks. The higher parts correspond largely with one of the largest of the numerous granite masses—probably of Caledonian age and hence known as the Newer Granites in contradistinction to the Older Granites more intimately associated with the Highland Schists now generally accepted as of pre-Cambrian age. The western part of the Grampian Highlands, though it has a similar summit level of 2000 to 3000 feet, may be described as a highly dissected plateau of which only peaks remain to indicate the height of the old peneplane. Extensive fault-belts, commonly trending north-east to south-west, have produced lines of weakness and tracts of shattered rock along which denudation has been relatively easy with the result that this part of the Grampians has many lochs—it is the region of the great lochs such as Tay, Fyne, Awe, Ericht, Rannoch, Katrine and Lomond. Although the eastern part of the Grampians has thus a gently rolling surface, contrasts are provided by the great corrie walls and by the deep forbidding glens which are deeply entrenched into its surface. The western part has, in Ben Nevis, the highest point in the British Isles (4406 feet) and it is in this area that the detailed lithology of the ancient rocks has played the major role in determining the form and distribution of mountain and valley. Hard quartzites make up Schiehallion, Ben-y-ghlo and the Paps of Jura, the metamorphosed Ben Ledi Grits form Ben Vorlich, Ben Ledi and Ben Lomond, whereas Ben Nevis is of granite. The valleys are generally excavated in the weaker limestones, slates and phyllites.

Westwards the country of fresh-water lochs passes into the fiord coast. The argument of the late Professor J. W. Gregory that the fiords were excavated along series of shatter-belts does not seem to be incompatible with the view that their final widening and deepening was due to tongues of Pleistocene ice. The north-eastern coastlands are very different—long straight stretches sometimes fringed with sand-dunes, including the Culbin Sandhills which have proved so difficult to fix and which by movement in the past have caused much damage. Inland is the Moray-Buchan platform, at about 500 feet above sea level. It is probably an old plane of marine denudation or platform on which Mesozoic rocks were deposited but have since been stripped away to re-expose the ancient floor. Over some parts of it, mainly to the west, Old Red Sandstone and some New Red Sandstone occur but over the larger part glacial drifts rest directly on the pre-Cambrian. It is this drift cover which provides soil adequate for extensive cultivation.

Although that remarkable trench, the Great Glen, has long been recognised as a feature dating from the far-off days of the Caledonian orogenesis (as its partial infilling by Old Red Sandstone renders obvious) it is only recently that evidence has been adduced to show the way in which it differs from the great thrust-faults which have been mapped in the Grampians and the North-West. It appears to be a line of movement along which there has been a huge *lateral* shift of the rocks—as it were a great tear in the earth's crust.

The North-West Highlands and Islands embrace the whole of Scotland north-west of the Great Glen. The oldest rocks are believed to be the Lewisian gneisses, so-called because of their widespread occurrence in the island of Lewis, though they form also the basal rocks of western Sutherland and Ross. They were intensely folded and metamorphosed in the earliest times and on their denuded surface were laid down, still in pre-Cambrian times, the massive red sandstones known as the Torridonian. Because of a well-marked vertical jointing they give rise to some spectacular cliff scenery, with great walls of sheer rock and sometimes fantastic pinnacles. An example which has been selected for illustration is Stac Polly in Wester Ross. On these Torridonian sandstones were laid down Cambrian strata, including a limestone, the Durness Limestone. This is interesting because though its outcrop is narrow it is sufficient to give rise, in this extreme north of Scotland, to a patch of typical limestone country, even recalling in several respects the downlands of southern England. But over these Cambrian sediments the great Caledonian earth-movements thrust a huge block of country—it was literally thrust or slid right over them and hence the narrow belt along which they are found to-day—consisting of metamorphic rocks which may be of the same age as those of the Grampians and give rise to scenery broadly like that of the western Grampians (see Fig. 73). Along the eastern coasts of Caithness, Sutherland and around Moray Firth as well as over most of the Orkneys are Old Red Sandstone rocks resting on the ancient gneisses—often flaggy in character but on lower ground covered with boulder clays which afford useful soils. Incidentally the Shetlands owe their great contrast to the Orkneys in that they are of the old gneissose complex.

Returning to the west coast, entirely different scenery is found in Skye and Mull and some of the smaller isles such as Rhum and Eigg. This was the great area of Tertiary volcanic activity—if the Alpine movements failed to fold again the old blocks they seem to have opened up huge fissures out of which poured vast sheets of very fluid basaltic lava ; extensive lava plateaus, nearly horizontal, are the result. Other cracks radiating right across to northern England were filled with lava which there consolidated to form wall-like dykes. From the large number the series are well named " dyke swarms." Some of the molten material was more sticky in character and consolidated to form irregular hill-masses of which the most notable are the Cuillin Hills of Skye. Of the lavas the vertical jointing of the basalt of famed Fingal's Cave in the Isle of Staffa rightly suggests an association with the lava of the same age which makes up Giant's Causeway in Northern Ireland. For a map of these Tertiary igneous rocks see Fig. 63.

IRELAND

ALTHOUGH the majority of examples in this book have been selected from different parts of England, Wales and Scotland, it will have become clear that the surface features of Ireland are closely connected in their underlying structure and geological origin with those of the larger island of Britain. Indeed, the major massifs of Highland Britain—such as the Highlands of Scotland, the Southern Uplands and the Welsh Massif—are actually continued into Ireland and the separation of the two main islands took place quite recently in terms of geological time.

One gets an immediate clue to the dominant factor in determining the present surface configuration of Ireland if one compares a solid geological map with a drift map of the island. The "solid" map—for example the general map of the British Isles on the scale of 25 miles to one inch—shows a vast stretch of Carboniferous Limestone covering the greater part of the heart of the island and reaching the sea in a number of places between masses, of varying shape and size, of older rocks which are arranged in a discontinuous ring. The central stretch of Carboniferous Limestone is also interrupted—in places by outcrops of older rocks which are thus clearly inliers, in places by outcrops of younger rocks forming outliers. A map of the drift deposits gives quite a different picture. A very useful map is the Map of Surface Geology of Ireland, by Sir Archibald Geikie, published by Bartholomew on the scale of 10 miles to one inch. It was prepared many years ago but remains to point the way to what England and Scotland have not got but badly need—a generalised drift map. This map of Ireland shows boulder clays occupying at least half the country as well as very large stretches of peat. Over the heart of the country the cover is so complete that the underlying Carboniferous Limestone, so widespread on the solid map, is rarely exposed at the surface. Even the upland masses of the rim are liberally plastered with drift. The drift map makes it clear, therefore, that practically the whole of Ireland was covered by ice at one time or another during the Great Ice Age. Several factors have combined to secure the maximum of interference

E. J. HOSKING

Millstone Grit scarp, north-west of Hebden Bridge, Yorkshire (See Fig. 27)

J. A. JENSON

Looking west to Crummock Water (See page 48)

J. FISHER

Stac an Armin, Stac Lee, and Boreray, St. Kilda, from the south-west (See page 53)

J. FISHER

Stac Polly from Loch Lurgainn, Wester Ross, capped by Torridonian Sandstone (See also Plate 1 and page 234)

with surface drainage. The centre of the island is a plain, or rather a low plateau, with a surface only about a hundred or a hundred and fifty feet above sea level, so that there is little natural fall. In the second place the rock underlying most of this plain is limestone and owing to solution of the rock and the development of underground watercourses the surface drainage is naturally irregular or ill-defined. In the third place this has been made far worse by glacial interference and the irregular deposition of masses of boulder clay, sands and gravels. In the fourth place the climate of Ireland is one of almost constant moisture which hinders the drying of the surface and promotes the growth of mosses and the development of bogs. Many of the great bogs in the central plain may represent the later phase of glacial lakes but the bogs of the west stretch over both hill and valley as typical " blanket bog " and as such are to be regarded as the climatic climax vegetation of the country. Whatever their cause the bogs act as a sponge, holding up moisture, and rendering still worse the already poor drainage.

The whole of Ireland is essentially a detached portion of Highland Britain. Apart from the large sheets of Tertiary basaltic lava in Antrim, the thin representatives of the chalk preserved under them, and the large Tertiary granite mass which forms the Mourne Mountains, practically the whole of Ireland is built up of Palaeozoic rocks—except, of course, for the Pleistocene drift deposits.

Broadly speaking, though the various rock groups and structural units may be traced across from Britain into Ireland, their distinctive features are softened, one might almost say, by the soft Irish climate, and sharp distinctions disappear. In the north the three major divisions of Scotland—the Highlands, the Central Lowland and the Southern Uplands—are continued across the narrow Irish Channel. The Highlands appear as the mountains and hills of County Kerry, of Donegal (where conditions most resemble those of the Scottish Highlands), of County Mayo and Connemara. Connemara is one of the most interesting regions of Ireland and is scarcely paralleled in Scotland—a glaciated rock surface almost at sea level with outcrops of bare rocks at intervals (see Plate XXIA) appearing through the blanket-bog and from which rise with startling abruptness the Twelve Bens or Twelve Pins of Connemara. The Southern Uplands of Scotland reappear as the uplands of County Down and County Armagh, complicated by the noble mass of the granitic Mourne Mountains and that

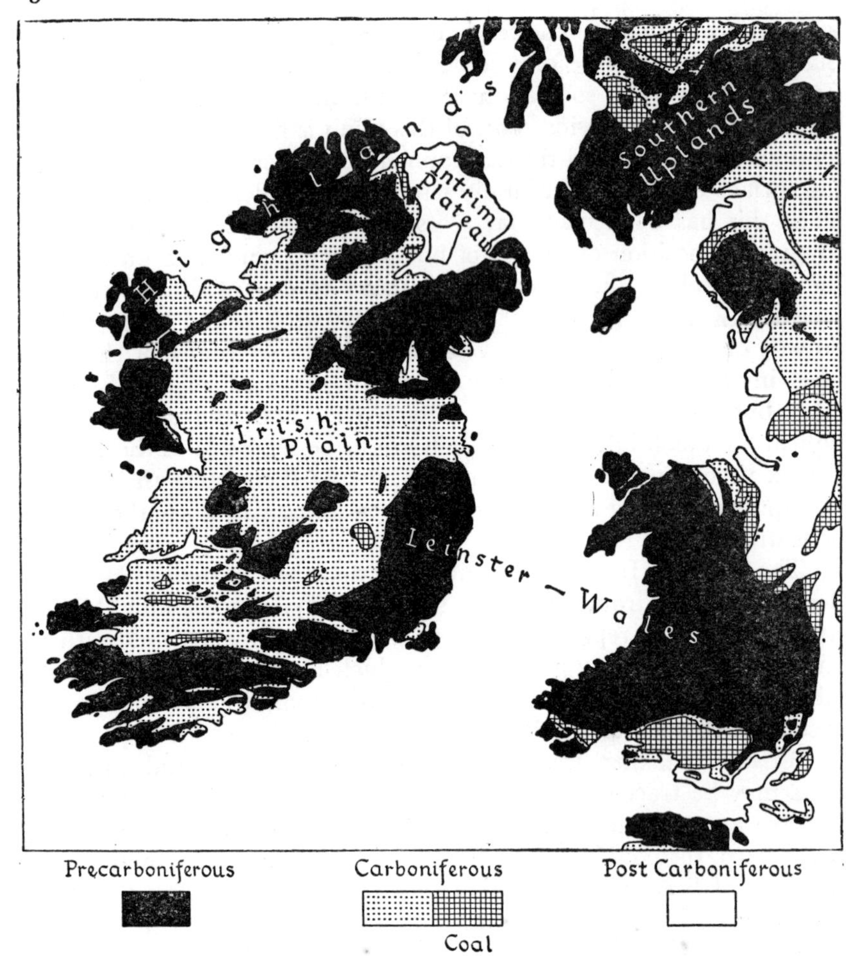

FIG. 74.—Diagrammatic map showing the structure of Ireland (after Mackinder)

majestic fiord, Carlingford Lough, where the " mountains of Mourne run down to the sea."

The Irish continuation of the Central Lowlands of Scotland is obscured for three reasons. The lowlands merge westwards into the central Irish plain, the boundary faults die out and the rift structure is thus less obvious—incidentally Ireland is unlucky in having no important coal seams in the rocks of Carboniferous age in this belt—but, even more important, the huge sheets of lava poured out in Miocene times stretch right across the area in Antrim. The eruption probably took place from hidden fissures and the lavas built up a great lava plateau which has collapsed in the middle and is there covered by the shallow waters of the largest lake in the British Isles—Lough Neagh. The basaltic lavas in cooling gave rise to the well-known hexagonal columns so well seen in Giant's Causeway but scenically the vertical jointing so caused has an even more important effect in providing the forbidding black vertical rock walls overlooking the Glens of Antrim. In those small areas where the chalk appears at the surface in Antrim it is interesting to record the spectacular reappearance of all the features normally associated with the downlands of southern England.

That part of Ireland which is now Northern Ireland has thus a varied and interesting scenic heritage. Eire has likewise its great contrasts between central plain and surrounding uplands. The central plain not only boasts the great bogs such as the Bog of Allen but many of its glacial hollows form lakes, most of them draining sluggishly to the Shannon. In outline these lakes are extraordinarily tortuous where the waters have invaded the boulder-clay hollows between drumlins and for the same reason they are studded with innumerable islands and islets. The same thing happens where the sea reaches the central plain, as in Clew Bay. The morainic ridges and much-discussed eskers (see page 87) form other features across the plain, but from the scenic point of view interest centres on the hill masses which rise from the rather monotonous level. Still greater contrast is offered where, especially over in the west, the Carboniferous Limestone emerges at the surface from under the mantle of boulder clay. There then reappear, notably on the borders of County Clare, the familiar limestone pavements of the English Pennines and associated sheep pastures.

Much of the central plain is drained by the Shannon and its tributaries and where the river enters Lough Derg it is a little over 100 feet above sea level. On both sides of the lake rise anticlinal masses of Old Red Sandstone and Silurian rocks and the waters of the Shannon leave the lake to pass by a series of rapids between Killaloe and Limerick over these old rocks. It is this which has made possible the development here of the Shannon Power works, generating enough electricity to supply the whole of Ireland. The southern mass of hills, the Silvermines and Slievefelim Mountains, are continued along the

Caledonian line north-eastwards into the Slieve Bloom Mountains, whilst to the south lies the celebrated Golden Valley, so-called because of its agricultural prosperity, which is actually a Carboniferous Limestone plain over which the drift deposits are well drained.

South-eastern Ireland, roughly the counties of Wexford and Wicklow, consists largely of Cambrian and Ordovician rocks : the strike lines are parallel to those of Wales and the area as a whole has a structural resemblance to central Wales, but the elevation of the land is less and this, combined with a light drift cover, relatively low rainfall and good drainage, results in considerable arable production including the growing of the wet-intolerant barley. There is another marked contrast with Wales in the great granite mass which makes up the Wicklow Mountains and their continuation south-westwards—the largest granite mass in the British Isles. Though not the highest mountains in Ireland, certain points of the Wicklow Mountains reach over 3000 feet ; they have the typical rounded forms associated with granite, whilst the deep secluded valleys on the east with their celebrated round towers have formed one of the cradles of Irish history from earliest times.

Though differing in many ways from South Wales, the south-west of Ireland is essentially Armorican and consists of long parallel ridges of Old Red Sandstone separating drift-filled valleys excavated in Carboniferous Limestone. The ridges and valleys trend west-south-westwards and where they run into the Atlantic Ocean they provide the classic example in these islands of an " Atlantic " type of coastline with drowned valleys or rias. This part of Ireland is also remarkable for the exceptional mildness of its winters so that, where there is shelter from the strong winds off the Atlantic, certain plants more typical of Mediterranean latitudes both survive and flourish. The Mountains of Kerry, developed on Old Red Sandstone, include the magnificent group of Macgillycuddy's Reeks with Carrantouhill reaching 3414 feet—the highest point in Ireland—overlooking the beautiful Lakes of Killarney on the Carboniferous Limestone. Immediately to the north-east is one of those areas which the geological map shows as a large tract of Coal Measures but the rocks resemble rather the Culm Measures of Devon and Cornwall in the absence or at least unimportance of the seams or streaks of coal.

THE END

ANNOTATED BIBLIOGRAPHY

THIS BOOK has attempted to trace the geological history of the British Isles and to show the way in which the wind and the rain, frost and snow and ice, streams and rivers and the sea are not only at work moulding its surface to-day but have been doing the same and in much the same way for countless ages past. Forces which have played a large part in many past eras—crustal movements, volcanic activity and glaciation for example—are less in evidence at the present day, but their work can be seen in the rocks of the earth's crust. Whilst many of the intimate details of scenery, such as the lay-out of the fields, the cultivation of the land and the disposition of the habitations of man, depend upon human activities, the broader features, such as the distribution of mountain and plain, of hill and valley, of river, lake and sea, are the direct reflection of the underlying geological structure and have been the special concern of this book. We have taken the picture a stage further by dealing briefly with the soil and so have sketched out the main features of the stage on which the drama of British natural history is played.

We have thus been concerned with certain sections of the field of geology proper—including historical geology or stratigraphy—and with what is usually called physical geography, and we have been specially concerned with the young science of geomorphology and a little with pedology. As so often happens with a field of study on the borderland of several well-established sciences, the literature is scattered and the studies are the partial concern of many scientific societies—hence the need to annotate this brief bibliography.

Publications of Scientific Societies and other corporate bodies.

The *Geological Society* (of London) may claim to be the premier geological society of the world and its *Quarterly Journal* includes the results of a large proportion of the original geological work carried out in this country but it has developed relatively little on the geomorphological side. The *Proceedings of the Geologists' Association*, an association specially of amateurs, have of recent years included many papers of direct relevance in the study of the relationship between geology and scenery as well as very useful summaries of the geology of different regions of the country—usually prepared as excursion handbooks and followed by reports of the excursions. The *Geological Magazine*, marked by an independence of outlook, has also published some articles, usually short, of importance. The *Geological Survey* publishes the official work carried out by its officers especially in the Sheet *Memoirs* issued in explanation of the published one-inch maps (see page 31) and reference is made below to the parts of *British Regional Geology*. Occasional papers of importance in the present connection appear in the *Geographical Journal*, the quarterly organ of the Royal Geographical Society and a few in *Geography* (the journal of the Geographical Association) and in the *Scottish Geographical Magazine* (Royal Scottish Geographical Society). Soil science is only indirectly concerned here; the journal to be noted is *Soil Science* though the reports of the Interational Congresses have perhaps more generally useful matter. From time to time important papers appear in journals which have a wider scope—such as the *Transactions of the Royal Society of Edinburgh*—as well as in the publications of local natural history and geological societies.

Physical Geology and Geography.

Arthur Holmes, Principles of Physical Geology, London, Nelson, 1944. New edition 1963. Modern and comprehensive work.

P. Lake, Physical Geology, 6th Edition, Cambridge University Press, 1958. A standard work, narrower in its scope than the last.

L. D. Stamp, Physical Geology and Geography, London, Longmans, 1938 and later editions. An elementary work but with many pictures and diagrams.

Historical Geology.

A. J. Jukes-Browne, The Building of the British Isles, London, 3rd Edition, 1911, Stanford. This was a pioneer work in tracing the geographical evolution of these islands.

L. J. Wills, The Physiographical Evolution of Britain, London, Arnold, 1929.

A. K. Wells, Outline of Geological History, London, Allen and Unwin, 1938. These two books use partly the same approach.

L. D. Stamp, An Introduction to Stratigraphy (British Isles), London, Allen and Unwin (Murby), 3rd Edition, 1957. This is a students' handbook combining the study of stratigraphical geology with tracing the evolution of geography.

A. J. Jukes-Browne, Students' Handbook of Stratigraphical Geology, London, Stanford, 1912. For long the standard work, now largely superseded by

J. W. Evans and C. J. Stubblefield (editors), A Handbook of the Geology of Great Britain, London, Allen and Unwin (Murby), 1930, and

J. K. Charlesworth, The Historical Geology of Ireland, 1963.

J. A. Steers, The Unstable Earth, London, Methuen, 1932.

Geomorphology.

J. Geikie, Earth Sculpture, London, 1904.

J. E. Marr, The Scientific Study of Scenery, London, Methuen, 6th Edition, 1920. Poineer works in the field.

Lord Avebury, The Scenery of England and Wales, London, 1888.

A. Geikie, The Scenery of Scotland, London, 1865. These two are now mainly of historic interest.

S. W. Wooldridge, and R. S. Morgan, The Physical Basis of Geography, An Outline of Geomorphology, London, Longmans, 2nd Edition, 1959. The standard British textbook, with bibliography.

C. A. Cotton, Geomorphology, Christchurch (New Zealand), Whitcombe and Tombs, 1942.

C. A. Cotton, Landscape, Cambridge University Press, 1941.

H. Baulig, The Changing Sea-Level, Pub. No. 3 Institute of British Geographers, London, Geo. Philip, 1935.

A. E. Trueman, The Scenery of England and Wales, London, Gollancz, 1938.

R. L. Sherlock, Man as a Geological Agent, London, Witherby, 1922.

Pedology (Soil Science).

S. G. Brade-Birks, Good Soil, London, English Universities Press, 1944.

G. W. Robinson, Mother Earth, Letters on Soil, London, Allen and Unwin (Murby).

G. W. Robinson, Soils, their Origin, Constitution and Classification, London, Allen and Unwin (Murby), 1932.

Pleistocene and Quaternary Geology and the Ice Age.

W. B. Wright, The Quaternary Ice Age, London, Macmillan, 2nd Edition, 1937.

P. G. H. Boswell, Presidential Address to Section C of the British Association, York, 1932.

L. J. Wills, Ditto, 1937.

H. Godwin, History of the British Flora, Cambridge University Press, 1956.

H. Godwin, Radiocarbon Dating and Quaternary History in Britain. *Proc. Royal Society*, B, 153, 1961, pp. 287-320.

L. D. Stamp, The Geographical Evolution of the North Sea Basin, *Journal du Conseil International pour l'Exploration de la Mer*, Copenhagen, 40, 1936.

W. B. R. King and K. P. Oakley, The Pleistocene Succession in the Lower Parts of the Thames Valley, *Proceedings of the Prehistoric Society*, 1936, pp. 52-76.

Regional Studies.

H. J. Mackinder, Britain and the British Seas, London, Heinemann, 1902, and Oxford, Clarendon Press, 1907. The classic work in which Mackinder introduces many of the fundamental ideas of regional division.

Jean Mitchell (editor) Great Britain: Geographical Essays. Cambridge University Press, 1962.

J. Wreford Watson and J. B. Sissons, The British Isles—A Systematic Geography, London, Nelson, 1964.

L. D. Stamp and S. H. Beaver, The British Isles, London, Longmans, 5th Edition, 1963. A standard geographical general survey.

A. G. Tansley, The British Islands and their Vegetation, Cambridge University Press, 1939. With important general introduction.

H.M. Geological Survey, British Regional Geology. This series of 18 paper-covered volumes, affords a well-illustrated and simply written account of the geology of Britain, region by region. Published before the war, stocks were destroyed by enemy action but revised editions have now appeared.

A. E. Trueman (editor), The Coalfields of Britain, London, 1954.

See also the volumes in the *New Naturalist*:

J. A. Steers, The Sea Coast, London, Collins, 1953.

S. W. Wooldridge and F. Goldring, The Weald, London, Collins, 1953.

L. A. Harvey and D. St. Leger-Gordon, Dartmoor, London, Collins, 1953.

Gordon Manley, Climate and the British Scene, London, Collins, 1952.

INDEX

Acheulian man, 161.
Acid soil, 97.
Aerobic organisms, 97.
Age of Earth, 15, 17.
Agricultural lime, 204.
Agriculture, Board of, 204.
Agriculture and Soils of Kent, Surrey and Sussex, 92.
Ailsa Craig, 25, 53.
Aire Gap, 225.
Alaska, 82.
Alde, R., 196.
Alder, 169.
Alder zones, 169.
Algae, calcareous, 111.
Allen, bog of, 239.
Alluvial cone, 46.
„ fan, 46.
„ flats, 48.
Alluvium, 41, 48.
Alpen im Eiszeitalter, 155.
Alpine, 19, 25.
„ folding, 139, 141.
„ storm, 28, 142, 144.
„ vulcanicity, 144.
Alps, 155.
Alum Bay, 182.
Amazon, Eocene, 141.
Ammonites, 26, 27, 28, 131.
Ammophila, 58.
Ampthill Clay, 133.
Anaerobic organisms, 97.
Analysis of soils, 92.
Ancaster Gap, 167.
Ancylus lake, 170-171.
Angle of rest of screes, 46.
Anglesey, 217.
Anglo-Franco-Belgian Basin, 141.
Anglo-Gallic Basin, 141.
Annan, R., 228.
Ant, R., 195.
Antarctic, 82, 121.
Antecedent drainage, 44.
Anthracite, 120.
Anticline, 42, 69, 70.
„ superimposed on Syncline, 145.
Antiquity of Man, 103.
Antrim, 78, 239
Ape Dale, 220.
Arctic fauna, 168.
" Arctic " flora, 165 *note*.
Ardennes, 111.
Arenig Grit, 109.
„ Group, 108.
Arêtes, 89, 221.
Arkell, W. J., 156, 158, 163.
Armagh, 237.
Armorican, 18, 25, 28.
„ folding, 211.
„ movements, 123.
Arran, 53, 80, 146.
Arrans, 108.
Artefacts, 153.
Artesian Basin, 179, 180.
Arthur's Seat, 24, 77.
" Artificials," 96.
Ash, volcanic, 79.
Ashdown Beds, 189.
„ Forest, 190.
Ashgill group, 109.
Ashtead Woods, 180.
Ashton-under-Lyne, 225.
Asymmetry of valleys, 69.
Atherfield clay, 181, 189, 190.
Atlantic climate, 157.
„ coast, 240.
„ coasts, 58, 61.
Atlas, National, 30.
Aureole, 25.
Aureoles, metamorphic, 78, 80.
Aurignacian man, 163, 165.
Aust Cliff, 129.
Avebury, Lord, 37.
Avon (Bristol), 38, 44, 140.
„ Gorge, 117.
„ Worcester, 163.
Awe, Loch, 233.
Axe Edge, 220.
Aymestry, 220.
„ Limestone, 109, 111, 220.
Ayrshire, 229.
„ basin, 231.
„ soils, 91.
Bagshot Beds, 143, 178, 180.
„ heath soils, 93.
„ Heaths, 180.
Bailey, Sir E. B., 138.
Bajocian, 133.
Bala, 108, 222.
Ballard Point, 181.
Banbury, 203.
Bars, 196.
Bartonian, 142.
Basalt, 24, 78.
Base-level, definition, 41.
Basement clay, 158.
Basins of deposition, Jurassic, 202.
Bass Rock, 77.
Batholiths, 78, 80.
Bathonian, 133.
Bath oolites, 75.
Battersea, 173.
Bays, 58, 61.
Beaches, formation of, 55.
„ raised, 54, 59.
„ shingle, 57, 58.
Bear, cave, 151.
Bedding planes, 62, 63.
Benches, wave-cut, 71.
Ben Ledi, 233.
„ Lomond, 233.
„ Nevis, 83, 89, 233.
„ Vorlich, 233.
Ben-y-ghlo, 233.
Bere, Forest of, 181.
Bergschrund, 87, 89.
Berkeley, Vale of, 207.
Berwickshire, 227.
Betchworth, 188.
Biotite, 16.
Birch-Pine zone, 168.
Birmingham, 208, 209.
Bituminous coal, 120.
Black Country, 120, 208.
Blackdown Hills, 65, 204.
Blackheath Beds, 178.
„ Pebble Beds, 45.
Black Mountains, 220.
Blakeney Downs, 195.
„ Point, 196.
Blanket bog, 84, 169, 237.
Blean Woods, 180.
" Blowing " of soils, 101, 195.
Blue Lias, 130.
Blyatt, A., 168, 170.
Bodmin, 214.

Bodmin Moor, 80, 81, 149, 211, 214.
Boggy moors, 210.
Bogs, 237.
„ cause of, 83.
Bolt Head, 212.
Bone-bed, 112.
Bone Bed, Rhaetic, 128, 129.
Boreal climate, 157, 168.
Borrowdale Volcanics, 108.
Boscombe sands, 143.
Bosses, 80.
Boswell, P. G. H., 156, 157, 163, 165.
Boulder-clay, 84, 154 *et seq.*
Bovey Tracey, 142, 215.
Bowland Fells, 225.
Boxstones, 151.
Boyn Hill Terrace, 50.
Brachiopods, 27, 28.
Bracklesham Beds, 143.
Brakes, 195.
Brampton, 87.
„ soils, 91.
Brandon, 76.
Breaks of Slope, 149.
Breccia, 22.
Breckland, 101, 195.
Bredon Hill, 65.
Breidden Hills, 221.
Brickearth, 50, 163, 164.
Bridlington, 158.
Bridport Sands, 203.
Brighton, 191.
Bristol Channel, 59.
Britain and the British Seas, 5.
British Association, 156.
British Regional Geology, 247.
Brittany, 19, 111.
Broads, 50, 100, 195-6
Bronze Age, 169.
Brown coal, 120.
Brown earths, 94.
Brown forest soils, 94.
Brown Willy, 214.
Brückner, E., 155.
Buchan Plateau, 234.
Buckland, W., 103.
Budleigh Salterton, 127, 213.
Bullhead Bed, 140.
Bunter Pebble Beds, 127.
Bunter sandstone geography, 126.
Bure, R., 194, 195.
Buried channel (Thames), 50.
" Burning " of soils, 95.
Buttes, 65.
Cader Idris, 21, 79, 89, 108, 221.
Caer Caradoc, 220.
Cainozoic, 17.
Cairngorms, 80, 233.
Cairnsmore, 228.
Cairntable, 231.
Caithness, 235.
Caithness flagstones, 63.
Caledonian, 18, 28.
„ folding, 108.
„ mountains, 32.
Caledonides, 111, 222, 227.
Camborne, 214.
Cambrian, 16, 17, 32.
„ geography, 105.
Cambridge, 204.
Camelford, 214.
Campsie Fells, 228, 229.
Cannock Chase, 209.
Cannon-shot gravels, 195.
Canterbury, 76, 180.
Caradoc group, 108.
Carboniferous, 16, 17.
„ Limestone geography, 114.
Carbo-Permian, 123.
Carlingford Loch, 237.
Carlisle, 225.
Carnmenellis, 214.
Carrantuohill, 240.
Carrick Roads, 60.
Carstone, 189.
Castleton, 74.
Cave animals, 162.
„ Bear, 162.
„ Man, 162.
Caverns, 73, 74.
Caves, 73.
Cementstone, 116.
„ Series, 116.
Cenomanian, 136.
„ Transgression, 138.
Cenozoic, 17.
Central Lowlands, 227 *et seq.*
Central Wales, 217.
Chalk, 14, 36, 62.
„ cliffs, 2.
„ formation of, 138.
„ Ireland, 239.
„ pipes, 76.
„ Rock, 188.
„ scenery, 75.
„ succession of, 188.
Chalking, 98, 204.
Chalklands, 177.
Chalky Boulder Clay, 160, 163, 194.
Charnian, 136.
Charnian axes, 133, 134, 140.
Charnwood Forest, 19, 38, 128, 136, 208.
Cheddar Gorge, 74, 117.
Cheesewring, 214.
Chellean man, 159.
Chelsea, 173.
Chert, 75.
„ definition, 190.
Cheshire coasts, 59.
Cheviot granite, 159.
„ Hills, 228.
Chilean deserts, 104.
„ nitrates, 98.
Chillesford River, 151, 152.
Chilterns, 65, 71, 150, 197, 204.
Chislehurst, 178.
Chronology, Glacial, 155, 157.
Chronology, post-Glacial, 167.
Cirque, formation of, 218.
Cirques, 48, 87, 88.
Clapham Common, 176.
Clare County, 239.
Clay-with-Flints, 140.
Clay-with-Flints, definition, 76.
Clay-Humus, 96.
Clay Vales, 197, 205.
Clayey soils, 95.
Claygate Beds, 178, 180.
Clays, 62.
Clee Hills, 220.
Clent Hills, 209.
Cleveland Hills, 131, 159, 203.
Clew Bay, 239.
Cliffs, 21.
„ erosion of, 57.
„ formation, 51 *et seq.*
Clifton Gorge, 44.
Climates of the past, 104.
Climatic-climax vegetation, 169.
Climatic cycles, 25.
Clints, definition of, 74.
Clwyd, Vale of, 221.
Clyde estuary, 59.
„ Firth of, 53.
„ R., 227, 231.
Coal, Kimeridge, 135.
„ origin of, 120.
Coalbrookdale, 209, 220.

Coalfields, 98.
Coal Measure climates, 121.
„ Measure geography, 122.
Coal Measure Vegetation, 104, 119.
Coal Measures, 19, 33.
Coast Erosion, 54-5, 196.
Coast Erosion, Royal Commission on, 54.
Coastal accretion, 55, 57.
Coastline, 2.
„ deposition, 54, 59.
„ drowned, 49.
„ erosion, 54, 59.
„ length, 2.
Coasts, 51 *et seq.*
Cobbett, W., 102, 197.
Coire, 89.
Collyweston, 63.
Column, geological, 12.
Commons, London, 176.
Concertina folding, 112.
Concordant summit levels, 221.
Conglomerate, 22.
Coniston Limestone, 222.
Connemara, 237.
Consequent rivers, 41, 43, 140.
Contact metamorphism, 24, 80.
Continental deposits, 142.
„ Drift, 28.
„ Shelf, 54.
„ slope, 61.
Contorted Drift, 194.
Coolins (Cuillins), 89, 146, 235.
"Coombe rock," 89, 163, 191.
Copper ores, 25, 212.
Coral limestones, 111.
Corallian, 133, 203.
Coralline Crag, 151.
Corals, 22.
Corbicula, 161.
Corfe Castle, 145, 183.
Corndon, 80, 221.
Cores, flint, 162.
Cornbrash, 133.
Cornlands, 195.
Cornstones, definition, 114.
Cornubia, 172.
Cornwall, 11, 33.
„ coasts, 59, 61.
„ granite, 25.
„ scenery, 150.
Correlation of strata, 11.
Corries, 87, 88.
Corve Dale, 220.
Corylus, 168.
Cotham marble, 129.
Cotswolds, 65, 68, 69, 75, 129, 201, 202.
Counties, 2.
Coventry, 208.
Crag-and-tail, 85, 90.
Crag beds, 151, 152.
Crags, 193.
Crags, cause of, 63.
Craven Fault, 224.
Crazy paving, 63.
Creswell, 165.
Creswellian man, 165.
Cretaceous geography, 137.
„ system, 14.
Crevasses, 87.
Criffel, 228.
Crinoids, 111.
Cromer, 151, 156.
„ Forest Bed, 151, 156
„ Ridge, 162, 194, 195.
Cross Fell, 224.
Crousa Common, 21.
Crumb structure of soil, 96.
Cuesta, 67, 190, 200.
Cuestas, definition, 68.
Cuillin (Coolin) Hills, 89, 146, 235.
Culbin Sands, 12, 101, 234.
Cultivated land, 7.
Culm Measures, 33, 240.
Cumbrian Dome, 223.
Current bedding, 63, 93.
Currents, longshore, 57.
„ ocean, 51.
Cut-off, 40, 48.
Cwm, 89.
Cycle of erosion, 70.
Cycles, 19, 26.
Cycles, climatic, 25.
Cycles of sedimentation, 144.

Dalbeattie, 228.
Danian, 139.
Dani-Glacial, 170.
Darby, H. C., 198.
Dart, R., 211.
Dartmoor, 21, 80, 81, 149, 211, 214, 215.
Darwin, C., 103.
Davies, D., 123.
Davis, W. M., 37, 45, 184.
De Greer, Baron, 167, 170.
Delabole, 212.
Delta, 48.
„ deposits, 41.
De Martonne, E., 37.
Denbigh, 221.
Dendritic drainage, 41.
Denes, 226.
Dent Fault, 224.
Denudation, 12.
„ cycle of, 22.
Deposition, 12.
Derbyshire caves, 163.
Derg, Lough, 239.
Derwent, R., 74, 205, 214.
Devil's Dyke, 89, 191.
„ Punchbowl, 190.
Devon, granite, 25.
„ scenery, 150.
„ south, 33.
Devonian, 13, 16, 17.
„ geography, 112
Devonshire lanes, 211.
Dibunophyllum, 116.
Diestian, 149.
Diluvium, 10.
Dip, definition of, 65.
„ regional, 140.
„ slope, 65.
„ valleys, 45.
Dirt-beds, 135.
Donegal, 237.
Dorking, 190, 191.
Dormant volcanoes, 25.
Dorset coasts, 61, 131, 202.
„ Somerset Basin, 131.
Dove, R., 74, 224.
Dovedale, 117.
Dover, 52, 192.
Dover, Strait of, 56, 149, 155, 157, 161, 169.
Down, County, 237.
Downham Market, 194.
Downland, 75.
Downs, 204.
Downs, North, 89, 177, 186, 187.
Downs, South, 89, 191.
Drainage, Irish, 237.
Drainage of Lowland Britain, 201.
Draining of the Fens, 198.
Drift, 155 *et seq.*
„ Contorted, 194.
„ deposits, 84.
„ geology, 30, 31.
Drowned valleys, 49, 50.
Drumlins, 86.
Dry gaps, 76.

Dry-point settlements, 173.
Dry Valleys of chalk, 76.
Dudley, 74, 109.
Dukeries, 209.
Dumfries, 228.
Dungeness, 56.
Dunstable Downs, 204.
Dunwich, 196.
Durham City, 226.
Durness Limestone, 105, 234.
Dykes, 24, 229.
Dyke swarm, 79, 144, 235.

Earth History, 18.
„ structure of, 22.
„ core, 22.
Earth's crust, thickness of, 22.
Earthquakes, 18.
East Anglia, 147, 193 *et seq.*
„ coast erosion, 196.
East Anglia, glaciology, 158.
East Devon, 213.
Eastern Glaciers, 194.
East Kent, 117.
„ Lothian, 227.
Eboracum, 210.
Ecology of Coal Measures, 123.
Eden Valley, 223, 224.
„ Valley soils, 92.
Edge Coal Group, 116.
Edinburgh Castle, 190.
Eigg, 235.
Eildon Hills, 228.
Elements of Geology, 103.
Elephas, 151, 157, 159, 161.
Ely, Isle of, 197.
Emanations, radioactive, 16.
Eoanthropus, 158, 184.
Eocene, 17.
„ Amazon, 141.
„ cycles, 142, 143.
„ geography, 140-1.
Eozoic, 17.
Epicontinental seas, 54.
Epping Forest, 177.
Epsom Downs, 149.
Eras, 14.
Ericht, Loch, 233.
Erith, 176.
Erosion cycle, 37.
Erratic blocks, 89.
Escarpment, 66, 79.
Eskers, 87, 161, 195.
Essex, 193 *et seq.*
Estuarine Beds, 133.
„ Series, 203.
Etna, Mt., 77.
Etruria Marls, 123.
Eustatic movements, 41, 147, 151, 214.
Eustatic movements, definition, 22.
Evolution, 28, 29.
„ geological, 3.
Exe, River, 5.
Exeter, 216.
Exfoliation, 19, 20.
Exmoor, 33, 80, 149, 211, 214.
Exmouth, 213.
Exposures, 13.
Extinction of species, 11.

Facies fossils, 29, 109.
Fairlight Clays, 189.
Fal, R., 49, 211.
Falmouth, 60.
False-bedding, 63.
Fan, outwash, 84, 86.
Faringdon, 137.
Farmyard manure, 96.
Farnham, 68.
Father of English Geology, 10.
Fault, 23.
„ displacement of, 23.
Faults, normal, 186.
„ strike, 186.
Fell Sandstones, 116, 225.
Felspar, 119.
Fenland, 158, 166.
Fenland soils, 101.
Fens, 193 *et seq.*, *esp.* 196.
Fertilisers, 96.
Field capacity of soil, 100.
Fife, 231.
Finchley, 86.
Findhorn, R., 49.
Fingal's Cave, 145, 235.
Fini-Glacial, 170.
Finland, 171.
Fire-clay, 120.
Firestone, 188.
Firn, 83.
Firth of Clyde, 53.
Fish, primitive, 26, 112.
Fish, Rhaetic, 128.
Fissure eruptions, 24.
Flagstones, 63, 113.
Flakes, flint, 162.
Flandrian, 171.
Fleet, R., 176.
Flint, 75, 221.
Flint implements, 162 *et seq.*
Flood, Noah's, 10, 103, 155.
„ plain, 41, 48, 165.
„ Plain Terrace, 50.
Floods, effects of, 22.
Florida swamps, 119.
Floristic provinces, 121.
Folkestone, 52, 192.
„ Beds, 189.
„ groynes at, 57.
Foraminifera, 138.
Forest of Dean, 33, 207, 219, 220.
Forest of Wyre, 209.
Forest Ridges, 184.
Forestry Commission, 195.
Formation, definition, 14.
Forth, R., 231.
Fossils, 27, 29.
„ definition, 10.
„ facies, 29.
"Fossil" landscapes, 38, 208.
Fowey, 211.
Fox, arctic, 168.
Foxmould, 1.
Freestone, 63, 203.
Frost, action of, 20, 83.
„ action on chalk, 205.
Fyne, Loch, 233.

Garvock, Hills, 229, 231.
Gas, natural, 186.
Gault, 136, 137, 188, 191, 203.
Geikie, Sir A., 30 *note*, 155, 236.
Geikie, J., 155.
Geoanticline, 145.
Geographical Evolution of Britain, 103 *et seq.*
Geographical Journal, 242.
Geological Column, 12.
„ Magazine, 241.
„ Society, 241.
„ Survey, 30, 31, 241.
Geological time, 13 *et seq.*
Geologists' Association, 157, 184, 241.
Geomorphological analysis, 154.
Geomorphology, 37.
Germano-British Sea, 125.
Geysers, 25.
Giant's Causeway, 24, 78, 145, 239.
Glacial Lakes, 161, 166, 167.
„ Period, 154 *et seq.*
„ Soils, 167.

Glaciated valleys, 87.
Glaciation, 82 *et seq.*
Glaciations, four-fold, 155.
Glaciers, 154 *et seq.*
Glamorgan, Vale of, 9, 217.
Glaslyn, R., 50.
Glen Coe, 82.
Glen Kiln, 108.
Glen Roy, 86.
Gley, 94.
Glinka, 92.
Globigerina ooze, 138.
Glossopteris, 121.
Gloucester, Vale of, 207, 220.
Gloucestershire Basin, 131, 202-3.
Goat Fell, 80, 146.
Godwin, H., 168-169.
Gog Magog Hills, 204.
"Golden Hoof," 99.
Golden Valley, 240.
Gondwanaland, 121.
Gorge, 49.
Goring, 47.
,, Gap, 177.
Gothi-Glacial, 170.
Gower, 74, 219, 221.
Graded Valley, 47.
"Grain" of country, 59, 215.
Grampian Highlands, 111, 231.
Grand Bank, 54.
Granite, 25, 78, 80.
,, weathering of, 81.
Graptolites, 27, 108.
Gravels, high-level, 150.
,, influence of, 180.
Gravesend, 176.
Gravity, action of, 20.
Grays, 176.
Great Dismal Swamp, 119.
,, Eastern Glacier, 159, 194.
Great Gable, 21.
,, Glen, 231, 234.
Great Ice Age, 36, 82 *et seq.*, 154 *et seq.*
Great Oolite, 133.
,, Orme, 117.
Greenland, 82.
Greensand, 14, 136.
Greensand soils, 187.
Greenwich, 176.
Gregory, J. W., 233.
Grikes, definition of, 74.
Grindstones, 117.
Grits, 119.
Groby, 208.
Groynes, 57.
Guildford, 68.
Gully erosion, 101.
Günz glaciation, 155, 157.
Gypsum, 128.

Hade, 23.
Hadrian's Wall, 79, 225.
Hall, Sir D., 92.
Haltwhistle, 225.
Hambledon Hills, 203, 205.
Hampshire Basin, 32, 140, 142, 181, 183.
Hampstead, 150, 176, 180.
Hangers, Beech, 177.
Hanging Valleys, 87, 223.
Hardness of water, 73.
Hard pan, 65, 93.
"Hard" rocks, 37.
Harlech Dome, 108, 217.
Hartfell, 108.
Hartz Mountains, 19.
Hassock, 189, 190.
Hastings Beds, 189.
Hatfield Moors, 167, 209.
Hazel, 168.
"Head," 89, 156.
Headlands, 61.
Headley Heath, 148, 149.
Headon Beds, 143.
Headward erosion, 44, 47.
Hearthstone, 188.
Heavy minerals, 149.
Hebrides, 32, 155.
Helford River, 211.
Helsby Hills, 209.
Helvellyn, 222.
Hemera, definition, 14.
Hengistbury Head, 143.
Hercynian, 18.
Hereford, Plain of, 220.
Herefordshire, red marls, 114.
Hessle Glacier, 195.
Hexham, 225.
High Cup Nick, 78.
Highgate, 150.
Highland Boundary Faults, 229.
Highland Britain, definition, 5.
Highland Schists, 233.
Highlands of Scotland, 32.
High Peak, 224.
,, Weald, 184, 186.
Hill-creep, 101.
Hindhead, 102, 190.
Hippopotamus, 157, 161.
Historical Geology, 104.
Hodnet Hills, 209.
Hog-back ridges, 200.
Hog's Back, 67, 68, 177, 186.
Holborn, 176.
Holmesdale, 191.
Holocene, 17.
Homesdale, 191.
Homo Neanderthalensis, 162.
Homo sapiens, 163.
Horse, 168.
Horsts, definition, 208.
Hot springs, 25.
Humber, Lake, 85, 166, 209.
Humber, R., 85, 166, 205.
Humus, 94, 96.
Hungry soils, 1, 95.
Hunstanton, 137, 165, 204.
Hurst Castle spit, 56, 57.
Hydraulic Limestones, 130, 202.
Hythe Beds, 189, 190.

Ice, action of, 20.
Ice Age, 154 *et seq.*
,, Ages, 25.
,, work of, 82 *et seq.*
Icebergs, 54, 158.
Ice-caps, 82 *et seq.*
,, British, 159.
Ice-front, 84.
Ice-sheets, 82 *et seq.*, 154 *et seq.*
Igneous activity, cycle, 25.
,, Rocks, 77 *et seq.*
Iguanodon, 135.
Immature soils, 94.
Impervious Rocks, 68.
Implements, Sub-Crag, 151.
Improved land, 7.
Inchcape Rock, 77.
Incised meanders, 49.
Index-fossil, 29.
Inferior oolite, 133.
Ingleborough, 224.
Insolation, 19.
Inter-glacials, 158 *et seq.*
Interlocking spurs, 46.
Interstadial periods, 163.
Introduction to Stratigraphy, 105.
Ipswich, 140.
Ireland, 236 *et seq.*
,, Carboniferous of, 117.
Ireland, maps, 30 *note*, 236.
,, separation of, 155.
,, south-west coast, 61.

Ireland, structure, 236.
Irish Sea Ice, 162.
Iron Age, 169.
„ ore, 203.
Ironbridge, 167, 221.
Ironpan, 65.
Isle of Ely, 197.
„ of Man, 32.
„ of Wight, 52.
Isostatic movements, 28.
Istria, 74.
Iver-Boyn Terrace, 161.

Jennings, J. N., 196.
Jerboa, 168.
Jointing, effect of, 68, 234.
Joint-planes, definition, 63.
Joints, 63.
Jura, 233.
Jurassic, 16, 17, 32, 138.
„ geography, 129.
„ geography (middle), 132.
Jurassic geography (upper), 134.
Jurassic scarps, 66, 70.
„ sequence, variable, 201.

Kainozoic, 17.
Kames, 87, 161, 195.
Karst, 74.
Katrine, Loch, 233.
Keele Beds (Staffs), 123.
Kellaways Rock, 133.
Kent Coalfield, 186.
Kentish Rag, 189, 190.
Kerry, 237.
„ Mts., 240.
Kettle-moraine, 86, 87, 161.
Keuper, 126, 128.
„ geography, 127.
„ Lake, 208.
Killaloe, 239.
Killarney Lakes, 240.
Kilpatrick Hills, 231.
Kimeridge Clay, 134, 203.
Kimeridge Coal, 135.
Kincardineshire, 108.
Kinderscout Grit, 119.
King, W. B. R., 156.
King's College, 105.
Knickpoints, 49, 50.
Knipe, The, 228.
Kop, Kopje, 65.

Laccolites, 79, 231.
„ definition, 80.
Laindon Hills, 180.
Lake deposits, 41.
„ District, 21, 22, 32, 74, 222 *et seq.*
Lakeland, 6.
Lakes, glacial, 82.
Lammermuir Hills, 227.
Lampreys, 112.
Lancashire Coalfield, 9, 215.
Land classification map, 30.
„ Forms, 37 *et seq.*
„ ultilisation map, 30.
Landenian, 142.
Land's End, 53, 214.
„ „ Granite, 80, 81.
Landslides, 63.
Langdale Pikes, 222.
Lapworth, C., 167.
„ Lake, 167, 209.
Lava, 11.
Lavas, interbedded, 79.
Law of Strata identified by fossils, 11.
Law of Superposition, 11.
Lawley, 220.
Laws, 77.
Leaching, 94, 195.
„ definition, 93.
Lead Hills, 227.
Leatherhead, 191.
Lea Valley, 165, 177.
Ledian, 142.
Leeds, 225.
Leicestershire Coalfield, 208.
Leith Hill, 187, 190.
Lenham Beds, 148.
Les Noires Mottes, 149.
Level, changes of, 148 *et seq.*, 156 *et seq.*
Lewis, 234.
Lewis, W. V., 37, 57, 89, 218.
Lewisian, 19, 32, 234.
Lias, 16, 36.
Liassic geography, 129, 130.
Lickey Hills, 208.
Lignite, 120.
Limerick, 239.
Limestone Country, 73 *et seq.*
Limestones, conditions of formation, 41.
Limestones, impure, 75.
Limestone soils, 94.
„ solution of, 73.
Liming, 98.
Lincoln Cliff, 65.
Lincolnshire, 131.
Lincolnshire, Limestone, 203.
„ Wolds, 159, 194.
Linton, D. L., 37.
Lithology, 37.
Little Eastern Glacier, 162, 194.
Little Ouse, R., 194.
Littorina Sea, 170-1.
Liverpool, 209.
Lizard, The, 21, 212, 214.
Llanberis, 221.
Llandeilo, 108.
Llandovery, 222.
„ Series, 109.
Llanvirn group, 108.
Loams, 95.
Loch Doon, 228.
Loess, 163.
Lomond, Loch, 233.
Londinium, 173.
London, 50.
„ Basin, 32, 140, 142, 173 *et seq.*
London Bridge, 173.
„ Clay, 142, 178, 180, 193.
London Palaeozoic ridge, 138.
London, site of, 173.
„ spread of, 176.
Long Mountain, 221.
Longmynd, 220.
Lothians, 116.
Low Weald, 184.
Lower Greensand, 14, 136, 187.
Lower London Tertiaries, 178.
Lowland Britain (definition), 5.
Lowthers, 227.
Ludlovian, 109.
Ludlow, 222.
„ Bone Bed, 112.
„ Series, 109.
Lutetian, 142.
Lydford Gorge, 215.
Lyell, Charles, 16, 103, 119.

Mackinder, Sir Halford, 5, 237.
Magdalenian man, 165.
Magillycuddy's Reeks, 240.
Magma, definition, 25.
Magnesian Limestone, 33, 125, 225.
Maidstone, 190.

Malham Tarn, 117.
Malmstone, 188, 191, 204.
Malvern Hills, 19, 33, 123, 220.
Mammal, earliest, 129.
Mammoth, 151, 157, 163.
Man, Isle of, 108, 224.
Manifold, R., 224.
Manx Slates, 108, 224.
Maps, 30 *et seq.*
„ geological, 30 *et seq.*
Marble, 24.
Marine bands, 13, 120, 123.
Market Weighton, 130, 132, 133.
Market Weighton axis, 202.
Markfield, 208.
Marling, 98.
Marls, Old Red, 113.
Marlstone, 131, 207.
Marmot, 168.
Marr, J. E., 37.
Marram grass, 58.
Martinique, 24.
Master joints, 63.
Mauchline Sandstones, 229.
Maximum glaciation, 160.
Mayo, 237.
Meander, 40, 48.
Meander-core, 49.
Meanders, incised, 49.
Medieval Fenland, 198.
Melbourn Rock, 188.
Melrose, 228.
Melt-waters, 84.
Mendip axis, 131, 133, 202, 203.
Mendip Hills, 19, 74, 117.
Mercian Highlands, 123.
Meres, 85, 161, 195.
Mersey, R., 207.
Mesas, 64, 200.
Mesolithic, 168.
Mesozoic, 17.
Metamorphic aureole, 25, 78, 80.
Metamorphism, definition, 24.
Methven, 229.
Microclimates, 1.
Microlestes, 129.
Midland Gate, 207.
Midlands of England, 33, 72, 207 *et seq.*
Midlands, glaciology, 156.
Midlothian, 227.
„ basin, 231.
Migration, alongshore, 106.
Miller, A. A., 37.
„ Hugh, 10.
Millstone Grit geography, 117.
Millstones, 117.
Mindel glaciation, 155, 157.
Miocene, 17.
„ geography, 145, 147.
Miocene leaf beds, 142.
Moffat Water, 228.
Moir, Reid, 151.
Mole, R., 191.
Mona, 217.
Monadnocks, 54.
Monazite, 149.
Monsoon, 119.
Montian, 139.
Mont Pelé, 24.
Moorfoot Hills, 227.
Moorland, 7; map, 9.
Moor-log, 168, 171.
Moraines, 84 *et seq.*, 161.
Morainic bar, 49.
Moray-Buchan Plateau, 234.
Moray Firth, 131, 158, 235.
Morayshire, 101.
Morecambe Bay, 48, 59, 223.
Moreton-hampstead, 214.
Moreton-in-Marsh, 159.
Mosses, 209.
Mottled Sandstones, 127.
Mount Sorrel, 208.
Mountain-building, 11, 18.
„ „ movements, 18.
Mountain Limestone, 116.
„ torrents, 46.
Mourne Mts., 80, 146, 237.
Mousterian, 162, 163.
Mull, Isle of, 79, 142, 144, 235.
Mull leaf-beds, 142.
Multiple use of land, 225.
Murchison, R. I., 15.
Muschelkalk, 126.
Musk-ox, 151.
Mynydd Eppynt, 220.

Nappes, 111.
Nar, R., 194.
National Atlas, 30.
„ Trust, 196, 198, 215.
Natural History of Selborne, 1.
Natural regions, 172.
Neagh, Lough, 78, 239.
Neanderthal Man, 162.
Necks, volcanic, 77.
Needles, The, 145, 181.
"Negative" areas, 225.
Neolithic, 165.
„ Forests, 196.
„ man, 168.
„ Raised Beach, 171.
Nesses, 196.
Nevé, 83.
Nevis, Ben (*see* Ben Nevis).
New Red Sandstone, 2, 25, 124, 224.
Newcastle, 225.
Newcastle-under-Lyme, 225.
Newer Drifts, 161 *et seq.*
„ Granites, 233.
Newfoundland, 54.
Newmarket, 193, 194.
„ Heath, 204.
Newton Abbot, 142.
Nile, R., 48, 161.
Nithsdale, 227, 228.
Nitrates in soil, 98.
Nitrogen-fixation, 98.
Norfolk, 36, 193 *et seq.*
„ Broads, 50, 100.
„ coast erosion, 56.
Normal fault, 23.
North Berwick Law, 77.
„ Downs, 177.
„ Sea, 161 *et seq.*
„ „ Drift, 158.
„ Wales, 217.
„ York Moors, 203, 206.
Northumberland Coalfield, 9.
North-West Highlands, 232-3.
Norway, 118.
Norwich, 194, 196.
„ Crag, 151.
Nottingham, 85, 209.
Nummulites, 26, 27.
Nunataks, 83.
Nuneaton Ridge, 208.

Oakley, K. P., 156.
Obsequent rivers, 44.
Ocean currents, 51.
„ troughs of past, 106, 108.
Ochil Hills, 228, 230.
Oilshales, 116.
Old Red Sandstone, 10, 13.
Old Red Sandstone geography, 112.
Older Drifts, 161 *et seq.*
„ Granites, 233.
Oldhaven Beds, 178.

Oligocene, 17, 142, 181.
Onion weathering, 19, 20, 21.
Oolites, 16, 200.
Oozes, 138.
Oppel, 131.
Ordnance Survey, 30 *et seq.*
Ordovician, 16, 17.
,, geography, 107.
,, volcanoes, 107.
Origin of Species, 103.
Orkneys, 158, 235.
,, separation of, 155.
Orogenesis, definition, 18.
Orogenic movements, 18.
Ostracoderms, 26, 112.
Ouse, R., 48, 166.
Outline of Historical Geology, 105.
Outwash fans, 210, 213.
,, Fan, definition, 84, 86.
Overflow channels, 85, 160 *et seq.*
Overfolds, 111.
Oxbow Lake, 40, 48.
Oxford, 159.
,, Axis, 133, 202, 203.
,, clay, 133, 203.
Oxfordshire, 131.
Oxidation in soils, 97.

Pacific Coasts, 59.
Palaeozoic, 17.
,, Platform, 112.
Paludina, 135.
Panning, 149.
Paps of Jura, 233.
Parish boundaries, 1.
Pavements, limestone, 74.
Peak, The, 224.
Peat deposits, pollen of, 157.
,, Fens, 198.
,, soils, 198.
Pedology, 91 *et seq.*
Pelé, Mont, 24.
Pembrokeshire, 219.
Penck, A., 155.
Pendleside phase, 224.
,, series, 117.
Peneplains, 22.
Peneplanation, 52, 65, 70, 215.
,, Cretaceous, 138.
Peneplanation, sub-aerial, 40.
Peneplanation, Wales, 153.
Peneplanes, Pliocene, 148.
Peneplanes, Scotland, 228, 233.
Peneplanes, Wales, 221.
,, submarine, 148.
Pennant Grit, 219.
Pennines, 9, 33, 78, 123, 222 *et seq.*
Penrith Sandstones, 224.
Pentland Hills, 231.
Perched blocks, 89.
Pericline, 186.
,, definition, 140.
Period, definition, 14.
Permeability of soil, 100.
Permian, 16, 17, 25, 104.
,, geography, 124-5.
Permo-Carboniferous, 123.
Perranporth, 58.
Persia, 213.
Persian deserts, 104.
Perth, 229, 231.
Pervious Rocks, 68.
Pewsey, 181.
Phacolites, 79; definition, 80.
Phosphates in soil, 98.
Physical Basis of Geography, 37.
Physiographical Evolution of Britain, 103.
Pickering, Lake, 167, 205.
,, Vale of, 205.
Pillow lavas, 108.
Piltdown man, 158.
Pine-Hazel Zone, 168.
Pine Zone, 168.
Pipes, 148.
,, in chalk, 76.
Pisolites, 200.
Plains, 41.
,, of denudation, 41.
,, ,, deposition, 41.
Plants, primitive, 112.
Plateaus, 65.
Platform, 200 ft., 153.
,, 400 ft., 149, 151.
,, marine, 53.
,, Palaeozoic, 112, 180.
Platforms, 69.
,, Pliocene, 149, 214.
,, in south-west, 214.
Pleistocene, 17, 156 *et seq.*
Pleochroic halo, 16.
Pliocene, 17, 36.
,, geography, 147 *et seq.*
Ploughing, 101.
Plucking action of ice, 89.
Plymouth, 74, 212.
Podsol, 93.
Pollen analysis, 105, 154, 157.
Polyzoa, 151.
Pompeii, 11.
Ponder's End Stage, 165.
Ponding of water, 160.
Pontesford Hill, 220.
Porosity of soil, 100.
Porous rocks, 68.
Portland oolites, 75, 203.
,, stone, 135.
Portlandian, 135.
Portsdown Hills, 145.
Post-glacial, 171.
Posthumous folding, 136.
Pot-holes, 46.
Pratje, O., 170.
Pre-Alpine folding, 139.
,, Cambrian, 17.
,, Chellean, 158.
,, Glacial Raised Beach, 158.
Principles of Geology, 103.
Profiles of soil, 92, 94.
,, ,, valleys, 46.
Provinces, floristic, 121.
Pteraspis, 112.
Pterichthys, 27.
Pteridosperms, 121.
Purbeck Beds, 135, 136, 189.
,, Isle of, 36, 145, 147, 183.
Purbeck lagoon, 135.
,, marble, 135.
,, rocks, 203.
Purfleet, 176.
Purple Boulder Clay, 163.
Putney, 173.
Pyramidal peaks, 89.

Quantocks, 216.
Quartzite pebbles, 127.
Quaternary, 16, 17.
Quaternary Ice Age, 156.

Radial drainage, 223.
Radioactivity, 23.
,, and time, 16.
Radstock, 123.
Ragstone, 189.
Rain, action of, 20.
Rainfall, map, 7.
Raised Beaches, 54, 59, 156, 158.
Ranmore Common, 148.

Rapids, 166.
"Rascally heaths," 102.
Rayleigh Hills, 180.
Reading Beds, 178, 181.
Reclamation of land, 198.
Red beds, 124.
„ Chalk, 136, 137, 204.
„ Crag, 151.
„ Devon, 213.
Redruth, 214.
Reed swamps, 100, 196.
Regional uplift, 200.
Regions of Britain, 172 *et seq.*
Reid, C., 59.
Reigate, 188, 190.
Rejuvenation, 41, 47, 49.
Relief of ground, 8.
Rendzinas, 94.
Reptiles, Jurassic, 28, 135.
„ Triassic, 128.
Rhaetian Alps, 128.
Rhaetic, 16, 17, 207.
„ geography, 128.
Rhine, R., 151, 152, 156, 161, 197.
Rhinoceros, 151, 157, 161.
Rhobell Fawr, 108.
Rhynia, 112.
Rhynie, 112.
Rias, 49, 61, 215, 240.
Rift Valley, 112, 228, 229.
Rim of S. Wales Coalfield, 117.
Riss glaciation, 155, 157.
River capture, 44, 201.
„ Drift Man, 161.
Rivers, action of, 20.
„ classification, 41, 44.
„ drift of, 71.
„ underground, 73.
„ under-ice, 87.
„ work of, 39 *et seq.*
Roches moutonnées, 85, 90, 221.
Rockall, 53.
Rock flows, 21.
Rocks, Sedimentary, 62 *et seq.*
Roman Bank, 198.
Romano-British Times, 198.
Romney Marsh, 169.
Ross, 234.
Rossendale Fells, 225.
Rothbury Forest, 116.
Rough Rock, 119.
„ Tor, 214.
Rowley Regis, 208.
Royal Society of Edinburgh, 242.
Royston, 204.
Rum, 235.
Running water, 39.
Russell, Sir E. John, 92.
Rusthall Common, 190.

Sahara, 104.
St. Austell, 214.
„ Bees Head, 224.
„ „ Sandstone, 224.
„ Breock, 214.
„ George's Land, 119, 120.
„ Kilda, 53.
„ Paul's Cray, 178.
Salisbury Cathedral, 135.
„ Craigs, 79.
Salpausselka, 171.
Salt, 128.
„ marshes, 59.
Sandbanks, 39.
Sand dunes, 12.
Sandgate Beds, 189.
Sand grains, 69.
Sandringham, 194.
Sandstones, 62.
Sandy soils, 95.
Sanquhar, 228.
Saunton Sands, 58.
Scandinavian ice, 157, 158.
„ Ice Sheet, 82.
Scarplands, 35, 66, 200 *et seq.*
Scarp-slope, 65.
Scarps, form of, 69.
„ stepped, 69.
Scenery, 37 *et seq.*
Scenery of England and Wales, 37.
Schiehallion, 233.
Scientific Study of Scenery, 37.
Scilly Isles, 53, 80, 162, 214.
Scolt Head, 196.
Scotland, 227 *et seq.*
„ Central Lowlands, 33.
Scottish Geographical Magazine, 202, 242.
Scottish-Scandinavian Continent, 114-5.
Screes, 20, 21, 46.
„ angle of rest, 46.
Sea, action of, 20.
„ work of the, 51 *et seq.*
Sea-cliffs, origin of, 150.
„ mats, 151.
Sedimentary rocks, 62 *et seq.*
Sedimentary rocks, definition, 11.
Sedimentation, cycles of, 22.
Sediments, 62 *et seq.*
Seend, 137.
Selborne, 1.
Senonian, 136.
Sernander, R., 168, 170.
Serpentine, 214.
Settlement sites, Thames, 176.
Severn Gorge, 167, 220.
„ R., 129, 140, 163, 207.
Sewerby, 158.
Shales, 62.
Shannon, R., 239.
Shap Fell granite, 89, 159.
Shatter-belts, 233.
Shelf, Continental, 54.
Sheringham, 194.
Sherwood Forest, 209.
Shetlands, 3, 235.
„ separation of, 155.
Shingle drifts, 55.
Shooter's Hill, 38, 150.
Shores, sandy, 58.
Shotover Hill, 203.
Shropshire, 220.
Sidlaw Hills, 229.
Sills, definition, 79.
Siltlands, 198.
Silts, 95.
„ Fenland, 198.
Siltstones, 116.
Silurian, 16, 17.
„ geography, 109.
„ Limestone, 208.
Siluro-Devonian folding, 111.
Silvermines Mts., 239.
Skiddavian, 222.
Skiddaw Slates, 108, 109, 222.
Skye, Isle of, 79, 89, 144, 146, 235.
Slate, 24, 63.
Sleaford, 167.
Slieve Bloom Mts., 240.
Slievefelim Mts., 239.
Slip faulting, 61, 192.
Slipping of cliffs, 52.
Sludge, 89.
„ glacial, 191.
Smith, William, 10, 204.

Snaefell, 224.
Snowdon, 69, 70, 79, 88, 89, 108, 217, 221.
" Soft " rocks, 38.
Soil creep, 20, 101.
Soil-drainage, 100.
„ erosion, 101.
„ groups, 92.
„ maps, 91.
„ phases, 95.
„ profile, definition, 92.
„ Science, 242.
„ series, definition, 95.
Soils, 91 *et seq.*
„ glacial, 167.
Solenopora, 111.
Solent, 181.
Solfataric stage, 25.
Solid geology, 31.
Solifluction, 20, 89.
Solway Firth, 59.
Solway Plain, 223.
Southampton Water, 181.
South Downs, 191.
„ Staffordshire Coalfield, 208.
South Wales, 218.
„ „ Coalfield, 218-9.
Southern Uplands, 227 *et seq.*
South-Western Basin, 127.
„ Peninsula, 211.
Spango, 228.
Speeton Clay, 135, 136, 204.
Sphagnum, 100.
Sphinx Rock, 21.
Spilsby Sandstone, 135, 136.
Spithead, 181.
Spits, 57, 196.
Spitsbergen, 121.
Sponge Gravel, 136.
Stable minerals, 96.
Stac Lee (St. Kilda), 53.
Stac Polly, 63, 234.
Stack, 53.
Stacks, 81.
Staddon Grits, 214.
Staffa, Isle of, 24, 78, 145, 235.
Staffordshire, 123.
Stainmoor, 225.
Stake, Hill of, 231.
Stalactites, 73.
Stalagmite, 74.
Start Point, 212.
Steers, J. A., 37, 196.
Steppe fauna, 168.
Stirling, 231.
„ Castle, 231.
Stonehaven, 229.
Stonesfield, 63.
Stow's Survey of London, 176.
Strait of Dover, 56, 149, 155, 157, 161, 169.
Strand, 176.
Stranraer, 228.
Strathmore, 229.
Stratigraphical break, 12.
Stratigraphy, 104.
Striding Edge, 21.
Strike, 44.
„ valleys, 43, 44, 45.
Stroud, 65.
Structural elements, main, 175.
Structure of soil, 96.
Submerged Forests, 59.
Submerged Forests, 59, 168, 196, 198, 215.
Subsequent rivers, 43, 44.
Suffolk, 36, 193 *et seq.*
Summit levels, concordant, 153.
Summit peneplane, 138.
Sunken roads, 191.
Superimposed drainage, 44.
Sussex marble, 189.
Sutherland, 234, 235.
Swallow-holes, 73, 191.
Swamps, cause of, 100.
Swanscombe, 161.
Sweden, chronology, 168.
Synchronism, 11.
Syncline over Anticline, 180.
Synclines, 69, 70.
Synclines, asymmetric, 71.
System, definition, 14.

Talweg, 46, 47, 49, 87.
Tal-y-llyn, 221.
Tamar, R., 49, 211.
Taplow Terrace, 50, 163, 164.
Tardenoisian, 168.
Tay, Loch, 233.
Tay, R., 231.
Tea-green marls, 128.
Teart pastures, 99.
Tectonic basins, 21 ; definition, 19.
Tees, River, 5.
Teign, R., 211, 215.
Teignmouth, 213.
Temple Church, 135.
Terminal moraine, definition, 84.
Terraces, river, 49.
Tertiary earth movements, 19.
Tertiary rocks, 32.
„ Volcanoes, 229.
Tethys, 142, 145.
Teviotdale, 228.
Texture of soils, 95.
Thames estuary, 59.
„ marshes, 176.
„ terraces, 71, 156, 159, 179.
Thames, R., 40.
„ „ terraces, 49, 50.
Thanet, 169.
„ Sands, 140, 178.
Thick Coal, 120, 208.
Thorne Moors, 167, 209.
Thorney, 173.
Thrust fault, 23, 232, 234.
Tidal movements, 51.
Tiger, 151.
Tilgate Stone, 189.
Till, R., 87.
Tilth, 96.
Time, geological, 13 *et sea.*
Time-scale, 14.
Tin ores, 25, 212.
Tinto Hill, 231.
Toad Rock, 190.
Toadstones, 117.
Torbay, 2, 215.
Torquay, 74, 212, 213.
Torridonian, 32, 234.
„ Sandstone, 63.
Tors, 21, 214.
„ origin of, 149.
Town and Country Planning, Ministry of, 30.
Trace elements, 99.
Traité de Géographie Physique, 37.
Traprain Law, 77.
Tree-ferns, 121.
Trent, 48.
„ Valley, 209.
Trias, 124.
Triassic, 16, 17, 104.
„ plain, 207.
Trilobites, 26, 111.
Truro River, 60.
Tunbridge Wells, 21, 190.
„ „ Sand, 189.
Tundra, 20.
„ fauna, 163, 168.
Turonian, 136.

Tweed, R., 228.
Twelve Bens (or Pins), 237.
Tyne Gap, 225.
Tyrol, 128.

Unconformity, definition, 12.
„ diagram, 24.
Underclay, 120.
Undercliff, 52.
Underground water, 68.
Upper Greensand, 188, 204.
U-shaped valleys, 87.

Valentian, 109-111.
Valley glaciers, 87.
„ profiles, 46.
„ sections, 46.
Valleys, shapes of, 39.
Varved sediments, 167.
Ventnor, 52.
Vesuvius, Mt., 17.
Vleis, 128.
Volcanic activity, 24, 77.
„ country, 77.
„ eruptions, 24.
Volcano, diagram of, 80.
Volcanoes, 11, 24.
„ ancient, 77.
„ dormant, 25.
V-shaped valleys, 46.

Wadhurst Clay, 189.
Walbrook, 173.
Wales, 217 *et seq.*
Warp, 49.
Warping, 49, 147, 151.
Warren, The, 192.
Wash, 48, 58, 158, 166.
Wash-outs, 121.
"Wastes," 7, 102.
Wastwater, 21.
Watcombe clays, 213.
Water, hardness, 73.
„ supply, 225.
„ „ of London, 176.
Water, underground, 68.
Waterfalls, 46, 87, 223.
Waterstones, 209.
„ Keuper, 128.
Water-table, 99.
Watford, 140.
Waveney, R., 167, 194.
Waves, action of, 20, 51.
„ 51; height of, 51.
"Weak" rocks, 65, 69.
Weald, 1, 36, 71, 137, 184 *et seq.*
Weald Clay, 189.
„ moors (Shropshire), 210.
Wealden Beds, 135.
„ Dome, 45, 142.
„ drainage, 43-45.
„ island, 141.
„ lake, 135.
„ Rivers, 43-45, 191.
Wear, R., 226.
Weathering, 9.
„ resistance to, 62.
Wegener, A., 28.
Wells, A. K., 105, 108.
Welsh Borderland, 24, 33, 74.
„ Cambrian, 105.
„ caves, 163, 165.
„ Ice Sheets, 158.
„ massif, 209, 217 *et seq.*
Wenlock, 222.
„ Edge, 74, 109, 220.
„ Limestone, 109.
Wenlockian, 109.
Wensum, R., 194.
West Cumberland Coalfield, 119, 223.
West Hoathly, 190.
Westminster, 173.
Westmorland Limestone, 223.
Wexford, 240.
Whin Sill, 79, 224.
White, Gilbert, 1, 191.
White, Lias, 130.
Wicken Fen, 198.
Wicklow, 240.
„ Mts., 240.
Wight, Isle of, 36, 52, 68, 145, 147, 181, 183.
Wills, L. J., 105, 156, 158.
Wimbledon, 176.
Wind, action of, 20.
„ gaps, 76, 85, 191.
Wind-waves, 51.
Windsor Castle, 145.
Winnatts, 74.
Wissey, R., 194.
Wolds, 197, 204.
„ Lincolnshire, 159, 194.
Wookey Hole, 74.
Wooldridge, S. W., 37, 148.
Woolhope Limestone, 111.
Woolwich, 176.
„ Beds, 178.
Wrekin, 38, 128, 208, 220.
Wren's Nest, 74, 109, 208.
Wright, W. B., 156.
Würm glaciation, 155, 157.

Yare, R., 194, 195, 196.
Yarmouth, 194.
Yes Tor, 214.
Yoldia Sea, 170.
Yoredale Phase, 117.
Yoredales, 224.
York, 82, 85, 210.
York, Vale of, 82, 159, 162, 207.
York moraine, 165.
Yorkshire Basin, 132.
„ Coalfield, 9, 225.
„ Ouse, 205.
Ypresian, 142.

Zonal index, 29.
Zone, definition, 14.
„ fossils, 29.
Zones, 29.
„ of pollen, 168-9.
Zuyder Zee, 198.